# How to Test a Time Machine

A practical guide to test architecture and automation

**Noemí Ferrera**

BIRMINGHAM—MUMBAI

# How to Test a Time Machine

Copyright © 2023 Packt Publishing

**Group Product Manager**: Rohit Rajkumar
**Publishing Product Manager**: Bhavya Rao
**Senior Content Development Editor**: Adrija Mitra
**Technical Editor**: Saurabh Kadave
**Copy Editor**: Safis Editing
**Project Coordinator**: Sonam Pandey
**Proofreader**: Safis Editing
**Indexer**: Manju Arasan
**Production Designer**: Ponraj Dhandapani
**Marketing Coordinator**: Rayyan Khan

First published: March 2023

Production reference: 2090823

Published by Packt Publishing Ltd.
Grosvenor House
11 St Paul's Square
Birmingham
B3 1RB

978-1-80181-702-8

www.packtpub.com

*For Hackerin, 2013-2022*

*For he was on my lap for the most part of this book.*

*May his loving soul rest in cat's heaven.*

*– Noemí Ferrera*

# Foreword

**Noemí Ferrera's** *How to Test a Time Machine* is a comprehensive guide to testing in the software development industry. This book is an excellent resource for anyone interested in software testing, whether you're an experienced professional or a beginner. (I mean, the book title alone should make you want to read it!)

Noemí's approach to testing is unique and refreshing, focusing on topics often overlooked in traditional testing literature. In this book, she covers topics such as continuous testing, cloud testing, AI for testing, mathematics and algorithms in testing, XR testing, and testing patterns, among others.

Testing is essential to software development, and ensuring that software is reliable, secure, and performs well is critical. As the software industry grows, so does the need for high-quality testing. *How to Test a Time Machine* provides a unique, comprehensive, and much-needed guide to testing modern software.

The book is divided into several chapters, each covering different testing-related topics. The first few chapters cover the basics of testing and the test pyramid. The middle chapters delve into more specific topics, such as UI automation patterns, continuous testing, and AI for testing. The latter chapters cover more challenging topics, including testing hard-to-test applications and taking your testing to the next level.

I've interviewed Noemí many times on my TestGuild podcast, and she's spoken at my online Automation Guild testing events, so I know she knows her stuff, and this book does not disappoint.

Noemí's writing style is engaging, making it easy to follow and understand complex topics. Her use of examples and practical advice makes this book a valuable resource for anyone interested in testing, whether they are beginners or experienced professionals.

In conclusion, Noemí Ferrera's *How to Test a Time Machine* (and *Other Hard-to-Test Applications*) is an excellent resource for anyone interested in software testing. Her unique approach and comprehensive coverage of various testing topics make this book a must-read for anyone looking to improve their testing skills and knowledge.

*Joe Colantonio,*

*Founder @TestGuild*

# Contributors

## About the author

**Noemí Ferrera** is a self-taught programmer and wrote her first comprehensive program at the age of nine. When she grew up, she proceeded to achieve a degree in computer science specializing in hardware in Spain, a bachelor's degree in software engineering in Ireland, and a master's degree in computer science in Spain.

She is an international speaker and participates in testing community conferences (such as Selenium, Appium, and Automation guilds) and engages with the community through Slack channels, Twitter, and her blog.

In the 2020 Test Guilds publication, she was named as one of the top 28 test engineers to follow, as well as one of the top 33 test automation leaders to follow in the 2019 Tech Beacon and as one of the 100 women in tech to follow and learn from by *agiletestindays.com* in 2023.

*There are so many people that influenced me in one way or another so that I could write this book, but I can't fit them all in here, so I need to summarize.*

*I would like to start with all my teachers that helped me enjoy writing, computer science, and English. However, the person that influenced me the most to learn and love this language has been my sister, Rosa. It's because of her influence that I decided I wanted to study abroad, which led me to work abroad, and to everything else that came after.*

*It feels like it was only yesterday that I joined the quality team at IBM, getting my first opportunity in the quality area. It would be too much for me to mention everyone I worked with, but know that you are also in my thoughts while I write these lines.*

*I also want to thank everyone in the testing community, especially those on the Selenium committee, who trusted and believed in me, even before I made any code commitment. Thank you, Maaret Pyhäjärvi, Lisa Crispin, Marit Van Dijik, Raluca Morariu, Sneha Viswalingam,Tristan Lombard, Diego Molina, Andrew Krug, Bill McGuee, Corina Pip, Manoj Kumar, Simon Stewart, and Marcus Merrell, for being inspiring, trusting, and welcoming. (And I'm sorry if I skipped you, there are so many people to thank here; you are still important!)*

*I really don't think I would ever have completed this book it if weren't for the Packt team. I would never have been happy enough with it. So a big thanks to eveyone in at Packt, especially Adrija and Bhavya, who cheered me up and kept me accountable.*

*I also wanted to thank Philip, the reviewer, who really took the book seriously and called out important problems with it, and Joe, who wrote the perfect foreword and made me tear up a little when reading it (who am I kidding? I was plain crying!).*

*I want to thank my friends and family for having the patience to deal with me while writing this book and when I told them I could not go out because I was behind on it.*

*Special thanks to my mom – she has always been there for me and her support and patience has been paramount.*

# About the reviewer

**Philip Daye** is a seasoned software quality professional with over 25 years of experience in the field. Currently the QA team lead at Insider Intelligence, he has a diverse background as a tester, manager, architect, and leader, and has worked with companies of all sizes to ensure the delivery of high-quality software. Philip is deeply committed to staying current with advances in the field and actively shares his knowledge and experience with others through speaking engagements at conferences and meetups, as well as by founding internal communities of practice.

*I'd like to thank my wife, Shirley, and sons, Simon and Zachary, who are always my biggest supporters. To Tariq King, who has been a true friend and mentor, for his encouragement to continue to learn and share. To Dionny Santiago, Justin Phillips, Patrick Alt, Keith Briggs, and David Adamo, who challenged me to bring my best every day. Finally, I'd like to thank the author for the opportunity to be a reviewer of this wonderful book.*

# Table of Contents

## 3

# The Secret Passages of the Test Pyramid – the Middle of the Pyramid 41

## 4

# The Secret Passages of the Test Pyramid – the Top of the Pyramid     63

# Part 2: Changing the Status – Tips for Better Quality

5

## Testing Automation Patterns    89

6

## Continuous Testing – CI/CD and Other DevOps Concepts You Should Know    131

# 7

# Mathematics and Algorithms in Testing    155

# Part 3: Going to the Next Level – New Technologies and Inspiring Stories

## 8

10

## Traveling Across Realities    253

11

## How to Test a Time Machine (and Other Hard-to-Test Applications)    289

12

# Taking Your Testing to the Next Level                       325

# Appendix – Self-Assessment                                  345

# Index                                                        347

# Other Books You May Enjoy                                    358

# Preface

From simple websites to complex applications, delivering quality is crucial for achieving customer satisfaction. *How to Test a Time Machine* provides step-by-step explanations of essential concepts and practical examples to show you how you can leverage your company's test architecture from different standpoints in the development lifecycle.

You'll begin by determining the most effective system for measuring and improving the delivery of quality applications for your company and then learn about the test pyramid as you explore it in an innovative way. You'll also cover other testing topics, including cloud, AI, and VR for testing.

Complete with techniques, patterns, tools, and exercises, this book will help you enhance your understanding of the testing process. Regardless of your current role within development, you can use this book as a guide to learn all about test architecture and automation and become an expert and advocate for quality assurance.

By the end of this book, you'll be able to deliver high quality applications by implementing the best practices and testing methodologies included in the book.

## Who this book is for

This practical book is for test owners such as developers, managers, manual QAs, SDETS, team leads, and system engineers that wish to get started on or improve their current QA systems. Test owners looking for inspiration and out-of-the-box solutions for challenging issues will find this book beneficial.

## What this book covers

*Chapter 1*, *Introduction – Finding Your QA Level*, introduces the test architecture, analyzing how different projects can be built and how that can affect the test architecture. It also discusses ways of achieving and improving the test architecture, depending on the stage that a project is at, and identifies the different roles that can participate in quality, their responsibilities, and skills.

*Chapter 2*, *The Secret Passages of the Test Pyramid – the Base of the Pyramid*, reviews what type of tools, techniques, and testing are done in the base of the test pyramid, including the differences between coverage and good unit testing and mocking. It also shows a way of automating the base of the pyramid, making these tests easier and more fun to write.

*Chapter 3, The Secret Passages of the Test Pyramid – The Middle of the Pyramid*, reviews the faces of the middle part of the test pyramid, defining the differences between the backend tests and how they relate to eachother, along with exploring sending APIs, schema validation, sending messages, and shadow and performance testing.

*Chapter 4, The Secret Passages of the Test Pyramid – The Top of the Pyramid*, covers some of the main tests systems should have, going from the top of the pyramid to the bottom. It suggests some extra tips and projects you can use to improve your quality at the top of the test pyramid.

*Chapter 5, Testing Automation Patterns*, examines several models for UI automation test code writing to create better-designed frameworks. It visits different models, such as remote topologies, and deals with files with objects and screenshots, and reviews some ways of automating the repetitive parts of test code.

*Chapter 6, Continuous Testing – CI/CD and Other DevOps Concepts You Should Know*, introduces continuous testing, CI/CD, and other concepts related to DevOps. It discusses different types of continuous testing and tools for CI/CD, and provides a basic continuous testing example.

*Chapter 7, Mathematics and Algorithms in Testing*, highlights the importance of mathematics in testing, including algorithms that can be specifically useful. It reviews the importance of data science in testing and includes techniques for test case analysis, with an early approach to **artificial intelligence (AI)**.

*Chapter 8, Artificial Intelligence is the New Intelligence*, discusses the role of AI in testing, including why you should learn it, core concepts, AI for testing, and testing AI applications, with some specific projects and examples to get you started and excited about AI for testing.

*Chapter 9, Having Your Head up in the Clouds*, explores the cloud, including how it can be useful for testing, how to create tools to measure testing performance, how to test appropriately in the cloud, and its dangers.

*Chapter 10, Traveling Across Realities*, reviews the different concepts and related applications of XR, including VR and AR, and what particularities they have in terms of quality verification. It includes some tools for XR development and testing in XR and briefly covers the concept of the metaverse.

*Chapter 11, How to Test a Time Machine (and Other Hard-to-Test Applications)*, reviews tips and tricks for testing challenging applications, covering games, difficult applications, non-testing automation, and even what to do with impossible-to-test applications. It starts by discussing how to test a time machine (which gives this book its name) and how this could be extrapolated to any other app and concludes by talking about test architecture.

*Chapter 12, Taking Your Testing to the Next Level*, offers some tips on how to further your career, find topics that inspire you to reach the next level, and align them to your working and learning styles so that you can build your career plan and, finally, find time to work on it. It includes ways of helping others in your team grow and reasons for doing what you do and caring about quality, alongside the ethics related to testing and how these topics will apply in the future.

# To get the most out of this book

While this book is meant to get everyone started in all the discussed topics, having some knowledge of programming is preferable as it will help you get the best out of the practical examples.

On the other hand, if you have computing knowledge, but not so much testing knowledge, you may find some of the examples too easy for you. Feel free to take them to the next level and try out the more challenging aspects of what is discussed here.

**Software, programming languages, and libraries covered in the book:**

Java, junit-jupiter:5.7.2, Python, unittest, TypeScript, JavaScript, C#, JSON, CSV, RabbitMQ, Cypress, Selenium 4, AirtestProject/Poco, YAML, 3D.js, pandas, Unium, Unity, NumPy, scikit-learn, sklearn, gym, and TestNG.

*Additional instructions to set up each of these are provided within each chapter. Make sure you install the mentioned programs and libraries for each chapter.*

**If you are using the digital version of this book, we advise you to type the code yourself or access the code from the book's GitHub repository (a link is available in the next section). Doing so will help you avoid any potential errors related to the copying and pasting of code.**

> Disclaimer
> The author is not writing in representation of her company and no content in the book is related to any of her work within her company.

# Download the example code files

You can download the example code files for this book from GitHub at `https://github.com/PacktPublishing/How-to-Test-a-Time-Machine`. If there's an update to the code, it will be updated in the GitHub repository.

We also have other code bundles from our rich catalog of books and videos available at `https://github.com/PacktPublishing/`. Check them out!

# Download the color images

We also provide a PDF file that has color images of the screenshots and diagrams used in this book. You can download it here: `https://packt.link/gFY0Z`.

# Conventions used

There are a number of text conventions used throughout this book.

`Code in text`: Indicates code words in text, database table names, folder names, filenames, file extensions, pathnames, dummy URLs, user input, and Twitter handles. Here is an example: "Note that the following files can be located within the `Benchmark` folder in our GitHub's repo."

A block of code is set as follows:

```
package chapter2;
public class Calculator {
    public int add(int number1, int number2) {
        return number1 + number2;
    }
}
```

Any command-line input or output is written as follows:

```
java -cp (path to the lib folder where testng is)\lib\*;(path
to the bin folder)\bin org.testng.TestNG testngByGroup.xml
```

**Bold**: Indicates a new term, an important word, or words that you see onscreen. For instance, words in menus or dialog boxes appear in **bold**. Here is an example: "We need a system to keep all the code versions together. The place where code is kept is called a **code repository**."

> **Tips or Important Notes**
> Appear like this.

# Get in touch

Feedback from our readers is always welcome.

**General feedback**: If you have questions about any aspect of this book, email us at `customercare@packtpub.com` and mention the book title in the subject of your message.

**Errata**: Although we have taken every care to ensure the accuracy of our content, mistakes do happen. If you have found a mistake in this book, we would be grateful if you would report this to us. Please visit `www.packtpub.com/support/errata` and fill in the form.

**Piracy**: If you come across any illegal copies of our works in any form on the internet, we would be grateful if you would provide us with the location address or website name. Please contact us at `copyright@packt.com` with a link to the material.

**If you are interested in becoming an author**: If there is a topic that you have expertise in and you are interested in either writing or contributing to a book, please visit `authors.packtpub.com`.

## Share Your Thoughts

Once you've read *How to Test a Time Machine*, we'd love to hear your thoughts! Scan the QR code below to go straight to the Amazon review page for this book and share your feedback.

https://packt.link/r/1801817022

Your review is important to us and the tech community and will help us make sure we're delivering excellent quality content.

# Download a free PDF copy of this book

Thanks for purchasing this book!

Do you like to read on the go but are unable to carry your print books everywhere?

Is your eBook purchase not compatible with the device of your choice?

Don't worry, now with every Packt book you get a DRM-free PDF version of that book at no cost.

Read anywhere, any place, on any device. Search, copy, and paste code from your favorite technical books directly into your application.

The perks don't stop there, you can get exclusive access to discounts, newsletters, and great free content in your inbox daily

Follow these simple steps to get the benefits:

1. Scan the QR code or visit the link below

https://packt.link/free-ebook/9781801817028

2. Submit your proof of purchase
3. That's it! We'll send your free PDF and other benefits to your email directly

# Part 1
# Getting Started –
# Understanding Where You Are
# and Where You Want to Go

Each application is different, each company is structured in a different way and with different resources and budgets, and each one has a different maturity state. This does not mean they should aspire to go to the next state; maybe they are right where they should be. There is no one-size-fits-all solution. Think about quality architecture as a ladder where you want to focus on reaching the next step to the one you are on. Go one step at a time. This section is going to be the first few steps of that escalator, beginning with finding out where you stand and then reviewing all topics related to the test pyramid, focusing on the obscure things that other books might not tell you about.

This part comprises the following chapters:

# 1
# Introduction – Finding Your QA Level

Each application is different, and each company is structured differently, with different resources and budgets. There is no 'one size fits all' solution with respect to quality. Instead, we can think about test architecture as an escalator on which each time you want to focus on reaching the next step to the one you are on now. However, to understand how to get to the next level, you first need to understand the level you are currently on.

In this chapter we are going to cover the following main topics:

- Figuring out which quality assurance level your company has
- How to get to the next quality assurance level
- Identifying priorities, improvement areas, and skills needed to reach the next level

## Technical requirements

None! Take this chapter as your guiding star in testing and in this book. I hope that as you and your company progress in quality, the chapter will have a different meaning each time you read it.

## Tips for absolute beginners

Whether you are a beginner at testing or even if you have an intermediate or advanced level of knowledge in testing, there are many ways in which you can find inspiration and knowledge: reading books and articles, attending talks, meetups, and conferences, attending courses, doing projects, participating in hackathons, working on projects, or playing around.

The test community is very friendly, and there are always people willing to help others.

The tools are already around those who are really committed to learning; the greatest difficulty is picking which one to use at any given time and being honest with yourself about how much time you are committed to spending on that.

This book might include concepts that are obvious to me, but you might be unaware of them. If that is the case, feel free to use the help of the references or search for those concepts before continuing to read.

I hope this book helps you find your inspiration and interesting projects to try out. Keep in mind Mark Twain's famous words, "*The secret of getting ahead is getting started.*"

# Figuring out which quality assurance level your company has

Getting started with a company's testing requires making a series of decisions that could be hard to change later on. To be able to determine the best system for measuring and delivering quality for a company, you can start by asking some clarifying questions.

I also suggest you ask them whether you are about to join a new company or team to set the basis for your job and understand what to expect of the position and whether you will be able to grow professionally and help the company given its circumstances.

## What development process is being followed?

Understanding the development process will help you understand how to test it better. Although companies are tending to move towards agile (for the reasons I explain next), there might also be companies that are perfectly content to be working with the waterfall system.

### Waterfall

This sort of process is less common nowadays, but it is still in practice, for example, when systems are delivered in conjunction with hardware, or there are no expectations over new versions.

Testing a waterfall system usually implies managing a lengthy list of tests that are added up per feature and establishing planning for executing all those tests before launching the system to the user. We want to test everything that we can think of, and that can cover something that might already be covered with prior tests.

In these systems, it is especially expensive to find a bug in production as it might give a bad name to the product, could end up being dangerous for the end user, and might require building a new batch containing the product fix and having to re-deliver the product for free. You can see some examples of this in the *Further reading* section at the end of this chapter.

The tests are usually handled by one or more teams of experts, and they could be separated from the developers that built the deliverable (except for unit testing).

Although this is not so strict in a waterfall approach, starting testing early is always recommended. This is a lesson that we have learned from agile, as the earlier bugs are found, the easier they are to debug.

## *Agile*

Although delivering bugs to the customers is something that could damage a company's reputation and should be avoided at all costs, companies working in agile can afford to have minor bugs, provided they can be fixed quickly and seamlessly. It is important that the user has a platform to report bugs and issues if some of them reach the users.

Not delivering quickly enough could give advantages to competitors, lose clients' interest, or even make the product irrelevant to the market. Therefore, the most important thing is to be able to balance the quality of the product with the quality of delivery time. This could mean having fewer but smarter tests. Testing something that was tested before will slow down each release. Since we have more frequent releases this could add up time. That said, every time an issue reaches production, it is important to make sure there is a new test in place that will cover this issue next time, while thinking about other tests we could have missed.

API testing gains importance over end-to-end testing in this sort of system (more on this in *Chapter 4, The Secret Passages of the Test Pyramid – The Top of the Pyramid*). This does not mean that long end-to-end, exploratory, or regression tests should not be done, but we can make use of the API to cover some of them (as we will see in detail in *Chapter 4*). Visual testing is also gaining importance in these systems.

To achieve more testing, the focus turns to deployment, as we could use servers that are out of reach for the users to do more extensive testing before deploying to them.

The relationship between developers building the feature and developers that are building automation or other members of the test team should be tight in this sort of system to avoid test repetition.

## *Fake agile*

Many companies have moved from a waterfall approach to a fast waterfall that looks a lot like 'agile,' but it is not really 'agile.' In the process of the movement, we can see a lot of testing being carried over and trying to fit it into a smaller period for sign-off before release. These companies can be identified by having an impatient scrum expert or a project manager asking whether the test case was done and signed off every 5 minutes after the deployment had finished.

The best solution for this is to work alongside dev-ops or developers in charge of the deployments to set up a server system that would allow for more intensive, safer testing and figures out a way of cleaning the list of existing tests. In other words, we want to turn this into an agile system.

# What coding language to use?

Unit tests should generally live in the same project of the class they test, using the same programming language.

If there are no tests in place yet, the advice would be to implement them in the same programming language as the developers are using. The reason is that if you turn to combined engineering or shift left in the future, the developers can take ownership of the tests. Sometimes this is difficult because you find that they are using different languages for the frontend and backend.

If there are any tests in place, the best thing is to follow along with what is there unless there is a major reason to rewrite them.

If there are too many tests in place that are not well maintained, then there are two ways of approaching this; refactoring or writing from scratch. Depending on how badly or well-kept the current tests are and how reliable their executions are, it might be easier to do the former or the latter.Here is a graph to help you decide which test code programming language to use:

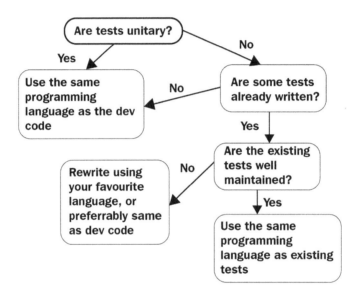

Figure 1.1: Identify the most suitable test code programming language for you

The way the code is shared and controlled also tells you a lot about the system required, as does the task management, bug tracking, and logging or dashboarding systems.

Finally, if technologies used for development are changing, it would be a good moment to change the testing approach, and it is also crucial to have stable tests to make sure these changes do not affect the users.

## What is the relationship between developers and testers?

As discussed in the previous sections, if developers and testers are well aligned, test repetition decreases, but independence might be affected.

### Do they sit together in the office?

Being physically near shows that the company cares for communication between development and testing. If the teams sit separately, this could start to form an *us and them mentality* that could be seen in the product afterward.

For companies that allow remote working, the question would be whether they are in different teams or under the same manager.

### Do they perform common code reviews?

Having a common code review system could result in finding bugs earlier. However, there is a general concern that having too many people signing off on a code review might slow it down or that the test team might not understand the code. I suggest that at least the unit tests are reviewed by someone with experience in testing, as code coverage tools could be easily deceived.

This also can have other benefits as all team members can improve their coding and testing skills.

## What is the size of the company?

Bigger and well-established companies can afford to have specific teams or people dedicated to tool creation, automation, and quality.

On the other hand, startups struggle most to budget dedicated experts and prefer to find people who can work in multiple areas and adapt to each situation. They might prefer to deliver more bugs in exchange for quicker proof of concepts and *faster* development. You could argue that developing bugs and having to spend time figuring out what is wrong and where, is not at all faster than developing with superior quality from the get-go, but some of these companies have few, if any, users and are working iteratively to find the right product rather than focusing on building a product right.

The right time for a startup or product to start integrating testing earlier is when the product is about to be launched to users or when the focus changes to building the product right. Unfortunately, by this time, it might be difficult for a tester to start working on automation, as they did not participate in the feature definition and development plan, which could result in testability problems. It will also be harder to write the unit tests then. Therefore, the advice is always to incorporate testing into the development, make it a habit, and get the right set of code reviews and structural automation in place. In this way, the time added will be unnoticeable. We will talk more about unit testing, testability, and the benefits of developing with a test mentality in the next chapter.

### Who writes unit tests?

This question might sound redundant, as nowadays **test-driven development** (**TDD**) is quite common, and most people understand the importance of unit testing and finding issues as early as possible. Nonetheless, this is not always granted, and some companies prefer someone other than the person that built the code to write all code for testing, including unit testing if needed. If that is the case, the test framework should include unit testing, and these tests might reside in a different place than the feature code.

### What are the key test roles?

Some companies might have **quality assurance (QA)** people, test automation roles, and tool creators, or it might all be done by the developers, with sometimes QA being done by the product owners.

Not all companies assign the name **software developer engineer in test** (**SDET**) or QA to the same role. Therefore, it is important that you understand what each role involves and what the expected responsibilities are to be able to help achieve better quality.

It is also important to understand the number of people that are working exclusively on quality versus development, to understand the amount of automation that will be needed.

### What does the app or company do?

This might seem obvious, but sometimes it is taken for granted. Depending on what the app or company does, you can think of the right and highest priority type of testing for it. For example, if the app deals with payments, localization and currency, testing is crucial.

### How many users do you have or plan to have?

The current versus potential growth is as important for testing as it is for development. Understanding the amount of testing that might be done and planning the best ways of debugging issues found during testing is not always done the same way for smaller projects as for bigger ones. Similarly, we should plan the right testing environments, which might need dedicated servers or other resources. There could be specific problems, such as multi-engagement across users, and performance testing could be of higher importance.

## Questions about the development

Having clarity about the way development is done in the company is highly important to create appropriate tests. We looked at development techniques before, but within these techniques, there could be differences in how the features are created. The following questions could also help clarify the quality process.

### How frequent are the project iterations?

The more frequent the iterations, the less time to perform testing. If iterations are very frequent and we find the need to do exhaustive testing, we need to be creative about it. For example, having some production servers disconnected for the users with the new code and applying testing there while other servers are being built before distributing them to the users.

*How long does it take to verify builds, and what is the goal?*

**Continuous integration/continuous delivery (CI/CD)** is the most automated way of doing this, but the tests could have different degrees of extensiveness depending on the goal.

*Is the priority aligned within the company? Per build or per feature?*

Is there a clear list of features per build? Is there a straightforward way of tracking which old builds could be affected by new ones? If not, regression tests will need to be more extensive (it is good practice to regression test everything but there is not always enough time for it).

As you can see, successfully defining the right quality process requires a lot of different inputs. There could be more parameters that could help gain a better understanding of the particular needs of a company or project for quality, but by now, you should have a better idea of what works best for each case. In the next section, we will see three examples of common traits that I have seen in the past and analyze how they could be improved. Additionally, you can see a self-assessment in the *Appendix*.

# How to get to the next QA level in a company:

In my experience, when someone asks for help getting started with testing, there are three statuses the company could be at, and therefore three things that might be needed:

1.  Convincing upper management of the need to test
2.  Convincing the other team members
3.  Getting a full framework up and running

## Convincing upper management of the need to test

As much as you are convinced that a particular set of tests are important if management believes otherwise, all testing efforts would land outside of expectations for the feature delivery, impeding the proper planning and delivery of tests. The first step before building a test framework or incorporating a new type of test is communication.

Managers' and product owners' main concerns usually center on time and budget. Addressing them is not always easy, and it highly depends on each case. The most important question is "why do we test?" or "why do we need this test in particular?" Make sure you can explain it and that you can add data to support the case.

## Convincing other team members

Similarly to convincing upper management, other team members might understand the need to test but not agree on where and how to do so.

If there is no test team in place, the best thing would be for the developers to start getting into testing by incorporating unit testing and TDD.

It is always nice to have someone with test experience to orchestrate tests, educate the team about testing, and start creating a test system for end-to-end and other sorts of testing or even perform some degree of manual checks. The same people or different ones could do this.

If the developers have never created any tests and/or are accustomed to other people verifying everything for them, it might be difficult to educate them on the use of unit testing and testability. But, as before, having enough data could help in the path of reasoning, even if that means you write some unit tests to prove their need.

If the issue is that they do not consider testing challenging enough, try to make them passionate about it. Maybe find some challenging test problems; there are some of them in this book.

Lastly, it might be that unit testing and mocking are taking as much time or longer than the time spent on developing the feature. If this is the case, consider automating some of it. (Yes! This is possible, and we will dive deeper into this in the next chapter.)

## Getting a full framework up and running

This is required when the company has some unit testing and/or manual testing in place but still does not have a test framework for a particular test type. Alternatively, they might have one but feel insecure about the current tests and want to make sure they are not missing anything to improve their system.

I will be explaining how to set some of them up throughout this book, but there is plenty of information online, comparisons, trials, and opinions, so do not feel obliged to set up the same frameworks that are here; pick the ones that suit your system the best.

# Identifying priorities, improvement areas, and skills

You should now have a better idea about the possibilities of testing around companies and applications. In this section, we are going to review some methods for improving each of the situations.

## Identifying improvement areas

The test mentality has a lot to do with asking questions and being curious. We started the chapter asking some questions, and here are some more to help you discover areas in which you could improve around quality:

Is there any repetition that could be reduced? If so, what? How could you automate those processes?

How many tests do you have of each type? We will be talking about the test pyramid in the following chapters; what does yours look like? Are there any other types of tests that could be beneficial for you? Are there any tests that are at the wrong pyramid level (or in more than one)?

How long does it take to sign off for each deployment? The ideal time should be under 15 minutes (which does not mean you cannot test further in other environments prior to or after deployment).

How much do you rely on your current tests? Are we testing what we should be testing? For example, I've frequently seen tests related to downloading documents from the browser. Unless your app is a browser, you should trust the download works properly. Is there another way of testing that document without going through the pain of downloading it? Are there other tests in which you are testing functionality that you do not need to test?

How much do you trust in your tests? If tests are not reliable or out of date, are they adding any value? Keeping the system clean is highly recommended. If you have a version control system, if a feature ever comes back, you should be able to retrieve the old test code from the version history.

Are you using any tools to track the issues and understand where we need or do not need tests?

Do you have the right tests for the right time? It is important to understand what tests to run and when to run each throughout development. We will discuss this further in the test pyramid chapters. We should also make sure we understand why we are testing or why we need a particular test. Avoid failing to do something just because other companies are doing it or imposing a set of tests that are not needed yet.

Lastly, if it is still hard to discern when something is needed, I highly recommend talking to a professional; some people would consider a short consultation for more tailored advice.

## Building the right team – testing roles and skills

Let us just take a bit of time defining testing roles as I have found that companies do not seem to agree about their definitions, and it is important to understand what people I am referring to throughout the book. I will also add some tips to help each of the roles grow in *Chapter 12, Taking Your Testing to the Next Level*.

Having a test expert help figure out the maturity of the company and what is needed to improve the quality of the product is particularly important. Test managers and test architects should be in distinct positions. However, not all companies need both positions, and sometimes, the job can be done by the same person. In some cases, automation is performed by developers, other times by developers in test, and QA, they are even called "automators" (which I believe to be a made-up word).

Rather than thinking of the following as "job positions," you could also consider them "roles" that can be performed by different professionals as needed.

### Test manager

A test manager makes sure the tests are performed correctly (created by the test or dev team) and good practices are implemented. They need to understand how and what to look for in the quality area. The position requires deep knowledge of testing and people skills.

## Test architect

The architect designs the frameworks and tools to be used for testing and can give practical development advice to people building test code. This position requires deep technical knowledge and experience in planning and coding tools from the start while having deep knowledge about testing. Sometimes this position is covered by an SDET.

## Software development engineer

Software development engineers (SDEs) are also known as developers. They are the people in charge of building features and, depending on the company, in charge of more or less of the test code.

## Manual testers

Some people refer to manual testers as QA testers. They are knowledgeable and passionate about applications, issues that could arise, and test methodologies. However, in some companies, QA testers also write some automation (generally for the **user interface** (**UI**)). Some companies invest in teaching automation, providing tooling to help them achieve more in less time, including the use of automated **behavior-driven development** (**BDD**) to turn the test definition into code, and visual testing in which UI screens are compared automatically with an expected image.

## SDET

SDETs are a rare species. They have the dichotomy of being developers with a testing mentality/passion.

Being stuck in writing test code could be frustrating for most developers as it is a repetitive task, not always challenging. When a company uses SDET to define the role that I am here referring to as QA, some people find themselves in that position, expected just to write automation code, have an unpleasant experience, and move away from the title.

Instead, they should be empowered to identify processes to automate and tools they could write to keep improving their programming skills.

Many companies are starting to join a movement called "shift left" or "combined engineering" in which SDE and SDET are combined into the "software engineer" role and work on all tasks related to coding, even test coding.

## DevOps

DevOps is a combination of developers and operations. A while back, tasks related to servers, deployments, and networks were done by a specialized team, sometimes referred to as "systems engineers." Most of the time, this also included a degree of system testing and even security.

As more programming languages, tools, technologies, and techniques started to develop, roles started to get increasingly specialized. However, the issues discovered by this team were usually difficult for them to fix, as they were not part of the development and needed an understanding of the code and features that they were not familiar with.

Here was where DevOps was introduced, in other words, developers doing this job for their team rather than across the company. In some companies, the "ops" bit is taken for granted and eliminated from the word, which I will do throughout the book.

### Other terms

Other terms related to testing are systems engineers or **system verification testers** (**SVTs**) (like "ops" but with more test knowledge and debugging capabilities), **functional verification testers** (**FVTs**) (a rather old term that involves end-to-end front and backend automation and testing), and integration testers.

For simplicity, they will all be referred to as SDETs in this book, even though there might be SDETs specialized in some areas as developers are.

## Scaling

Horizontal scaling means that you add more machines to your testing. For example, using the cloud to test in different systems (more web browsers, mobile versus desktop, and different operating systems).

Vertical scaling means that you scale by adding more types of tests (security, accessibility, and performance). We will talk about the "test pyramid," where each test falls on it, and what percentage of the time spent in testing should be done in each of them.

Identifying the type of testing you need and what systems should be covered should be part of an architecture design. We will talk about diverse types of testing and tools for horizontal scaling in *Chapter 2*, *Chapter 3*, and *Chapter 4*.

## Automating the automation

Time is the highest-valued currency you have; it is something you can never get back. This is one of the reasons why I like automation, so I can save time (mine or other people's). And by automation, I do not just mean test automation but any repetitive process. That is what I mean by "automating the automation."

In most companies, the "task automation experts" are the SDETs. That said, if you are a developer or a QA tester, you could highly benefit from this practice.

I have identified some basic steps for automating anything:

1.  **Recognizing automatable tasks**:

    The first step to automating something is to think about repetitive tasks. Then, you should consider how long it would take to automate that task and calculate how much time it would save if it were automated.

    You can tell when a company is doing well on their test automation when you see this thought process translated into it, rather than automating as many things as possible as some proof of skills or performance. The same concept can be extrapolated to any other repetitive tasks, including, of course, automation.

2.  **Write some code that does that task for you**:

    Once you have a clear picture of the steps involved in the repetitive task, you should also have an idea of how to automate those steps. The exact language, style, and patterns are up to you to define.

3.  **Identify when the code needs to be executed**:

    Sometimes we want to execute the code after something else happens. For example, testing automatically after a feature has been developed is common. Or, we can have automation that depends on some trigger, such as a page being refreshed or an application launched. Other times, we want the execution to happen at a certain point in time, for example, every morning. That could be automated with a cron job (also known as scheduled tasks) or services.

4.  **Identify success measures**:

    The next step is to identify our gain from this automation. What do we need to achieve with it? What is our best result metric? Sometimes, to verify that the automation has been executed and to check its success, we rely on logging. However, checking the logs could also be considered a manual task, and we will need to make sure we automate it if that is the case.

    I suggest creating an alert if something has gone wrong, for example, a test case failing. We may also have a notification that everything has gone well, too, just to verify it has worked. An email with the details, a text message, or even a phone call could all be ways of automating the logs (more details on notifications in *Chapter 11*, *How to Test a Time Machine (and Other Hard-to-Test Applications)*).

## *Where to start?*

When there are a lot of things a company or team could improve or even automate, I have a formula for automation that we can apply to other things:

(Time spent in doing the task manually + monetary value (based on the cost of potential issues)) / (time that will take to build the automation).

I argue that we could also think of it this way:

(Time spent in doing something before a change + monetary value (how much the task is needed)) / (time that will take to implement the change).

For "time spent," I mean total time, if there are three steps, we need to multiply it by the potential times those steps will be repeated. Of course, there might be other factors to consider as well, for example, the scalability of current versus future solutions and people impacted by this, but I found this to work for general cases and as a baseline.

## Moving on

Imagine that all the feasible options in this chapter are checked and implemented, and one or more test frameworks are set up. The team is managing the new tests, which are carefully planned and executed as part of the CI/CD structure. The analytics are set up and create alerts for any problems or improvements. So, can you ever be done with testing?

I presume most people in quality will jump straight away to this answer: "you are never done with testing." However, in the same way as you might "be done" with development, you need to "be done" with testing. It is normal if you are not 100% happy about your project, but you have done your best, and (usually due to budget) maintenance is all there is left, at least for that project (and until some cool tool or technology comes along).

What to do then? As with development, at this point, we can move on to the next project. Of course, someone should stay to fix potential issues and provide support, but you should not need such a big team or even the same skills for this. Not worse or better, simply different. Some people enjoy being at the beginning of the project, and others prefer maintaining them.

Development and testing go hand in hand; when there are fewer changes in development, there are fewer changes in test code.

## Summary

In this chapter, we had an introductory overview of the test architecture and analyzed how different projects could be built and how that could affect the test architecture. Since every company is different, we need to take into account all the possible inputs to produce the right set of testing. We can conclude that as development is intertwined with testing, decisions should be made for both of them at the same time.

Then we discussed ways of achieving and improving the test architecture depending on where the project is at that point in time. We considered what can be automated and how, and what process automation follows.

We identified the different roles that can participate in quality, their responsibilities, and their skills.

In the next chapter, we will discuss what is called "the test pyramid" and we will analyze its base. The base of the test pyramid is usually handled by developers, so they may be the most interested in this chapter, although SDETs and QA testers interested in growing in their careers might find automation that they could do in this section.

In the two chapters after the next, we will talk about the middle and the top of "the test pyramid," discussing in detail different test types and some tips and tricks around them.

## Further reading

Two examples of critical defects can be found with details at `https://cloudwars.co/ibm/bank-of-america-cloud-slash-billions-in-it-costs/` and `https://techtrends.com/2019/04/19/a-software-bug-caused-boeings-new-plane-to-crash-twice/`.

# 2

# The Secret Passages of the Test Pyramid – The Base of the Pyramid

Test strategies have been explored throughout the years, with the common goal of finding the right amount of testing. The questions of why we test, what we test, and more importantly when we stop testing have puzzled test experts for years, resulting in almost philosophical conversations.

Testing has been divided into many groups and categories, the most famous of them being **the test pyramid**. This term is quite interesting, as its graphical representation is that of a two-dimensional triangle rather than a tridimensional pyramid, shown as follows:

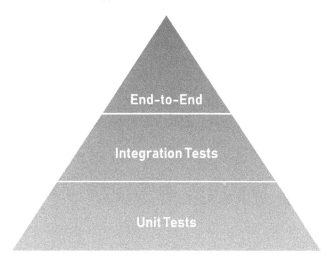

Figure 2.1: 2D representation of a test pyramid

In the following three chapters, we will take a different approach to the test pyramid, trying to bring this third dimension into the picture. We will also be observing the test types that usually are forgotten in this schema and alert you to the hidden dangers on our virtual pyramid.

In this chapter, we are going to cover the following main topics:

- What is inside the base of the pyramid?

- How to cheat on unit testing to achieve coverage and why you shouldn't

- Do not kill the mockingbird – how to level up your unit testing with mocking

- Secret passage – automation at the base of the pyramid

## Technical requirements

A basic level of programming skills is recommended to get the best out of this chapter. The last section might require an intermediate level.

Developers might be especially interested in this chapter.

This chapter uses examples written in various programming languages for everyone's convenience. We recommend reviewing and working with different languages as a self-growth exercise, and we provide the examples so it is easier for you to play with them in GitHub: `https://github.com/PacktPublishing/How-to-Test-a-Time-Machine/tree/main/Chapter02`.

In this chapter, we will use **Java** with `junit-jupiter:5.7.2`, **Python** with `unittest`, and **TypeScript** code.

## What is inside the base of the pyramid?

The base of the pyramid contains those tests that are closest to the development of the feature. Although there are some discrepancies around this claim, it is usually where most tests should take place. The reason is that the earlier the issues are found, the cheaper it is to fix them. In this case, the monetary value of finding an issue is measured by the amount of time spent figuring out why it has happened (debugging) plus the amount of time spent fixing it and the context switch needed if the developer has moved on to a different task.

Tests closest to the feature's development are sometimes referred to as "structural." In the 2D test pyramid (or triangle), the tests that are located here are **unit tests**.

Most developers include testing tools and processes in their development, although, interestingly, some of the tools and processes are not even considered as part of quality by them. Some developers claim that they "do not need testing" when they are, in fact, actually testing without realizing it.

## Unit testing

In unit testing, the focus is on testing the smaller units of code, which generally means testing each method. However, different tests can be performed on each of those units of code.

## Other dimensions

The other dimensions of the pyramid include tests and techniques to reduce the number of defects in the code, improve its testability, optimize the code, and verify missing lines or cases from tests, such as the ones covered in the following subsections.

### Code coverage

This is a technique to measure the lines of code that are traversed by the unit tests, associating this with the amount of code that is covered by tests. While this gives us an accurate idea of the coverage of the tests, this technique does not always give us an accurate idea of how well written those tests are. We will uncover issues with it in the next section of this chapter.

### Code optimization

Most people do not consider code optimization techniques part of testing. However, code optimization is a technique that looks for better code quality. There are a lot of automated tools, some of them even configurable, and techniques that could be used for that purpose.

### Code reviews

Compilation, code checkers (linters and style cops), and manual code reviews aim to increase the quality of the written code and therefore, should be considered part of testing. As they test the structure of the code, they should also be part of the bottom of the test pyramid.

Integrating a test expert (whichever role they might have) in the code reviews has some benefits, especially if the review includes unit testing and this expert is later writing test code. Repetition of tests across the pyramid could be an unnecessary, avoidable expense.

## The secret passages

Lastly, we could consider "mocking" (along with other related techniques such as "stubbing") our test passages, which allows the developer to ensure that the code is working with different outputs without expecting these to appear naturally.

We could argue that testability (writing clean code in a way that could be tested easily and avoiding code smells) could also be a "secret passage."

Database testing would be a passage into the next level of the pyramid. The testing of the databases is generally done structurally but usually after the feature code is tested.

# How to cheat on unit testing to achieve coverage and why you should not

Test coverage is a useful tool for making sure the unit tests are good enough. However, if they are used without being conscious of this fact, without good test knowledge, or without fully understanding why tests are needed in the first place, they will not be as useful as intended by their creators.

For example, imagine code for testing a calculator, as follows:

## Calculator.java

```
package chapter2;
public class Calculator {
    public int add(int number1, int number2) {
        return number1 + number2;
    }
}
```

To make the coverage pass with 100% accuracy, we simply need to make sure we call the add method with any two numbers.

The following is an example of such code:

## CalculatorTest.java

```
package chapter2;
import org.junit.jupiter.api.Test;
import static org.junit.jupiter.api.Assertions.*;
class CalculatorTest {
    @Test
    public void TestAdd() {
        Calculator calculator = new Calculator();
        assertEquals(7, calculator.add(3,4));
    }
}
```

However, unexpected errors can occur if we call this method passing two big numbers as parameters that add up to a number above the Java limit for storing an integer: 2,147,483,647.

If this code is part of a greater source, with many layers and integrations, we find ourselves in front of two difficult tasks:

- Figuring out where exactly the issue is

- Understanding what the issue is, in this case, an integer overflow

Therefore, code coverage cheating is a matter of making sure the basic tests are covered throughout the code base, including any potential branches. However, while code coverage gives us an idea of the amount of testing done for the unit of code, we could have 100% coverage but still have issues in the code, as we can see in the next example.

This is another possible implementation of a method to add two numbers for a calculator. In this case, we are adding a branch (with the if sentence) to print out a message if one of the numbers is negative:

## Calculator.java – method for adding non negative numbers

```java
public int addNonNegative(int number1, int number2) {
    if (number1 < 0 || number2 < 0) {
        System.out.println("Negative number");
    }
    return number1 + number2;
}
```

The following is an example of a test class for the preceding code, with a test per branch that is still missing an issue:

## CalculatorTest.java – test branches

```java
    /**
     * Tests adding two positive numbers
     */
    @Test
    public void TestAdd() {
        Calculator calculator = new Calculator();
        assertEquals(7, calculator.add(3,4));
    }

    /**
     *Tests adding a negative number and a positive one
     **/
```

```
@Test
public void TestAddNegative() {
    Calculator calculator = new Calculator();
    assertEquals(1, calculator.add(-3,4));
// ...
}
```

Note that the tests that would go on testing number 2 will be testing with the second input negative (example (3, -4)) and with both inputs being negative (example (-3, -4)), so we have omitted them as obvious.

Some strategies, such as **test-driven development** (**TDD**), suggest doing this exercise backward: first writing the code and then writing the feature. The added benefit of doing this in this way is that it should take fewer iterations than if the code needs to be fixed after the test fails.

Let's explore other tests to take into account during unit testing that might be not caught by code coverage. Following there is a list of them and also a list of potential checks that you can use in your testing. Feel free to use this as a checklist.

## Statement testing

In statement testing, each statement is executed at least once. But what if we forget an `else` to an `if` condition? We might have an apparently full test coverage and yet miss this issue.

For example:

---

### Calculator.java – method for subtracting

```java
public int subtract(int number1, int number2) {
int subs = 0;
if (number1 < 0 || number2 < 0) {
        System.out.println("Negative number");
        subs = number1 - number2;
        }
        return subs;
}
```

> **Note**
> Do not forget to check statements: verify the behavior upon entering the condition (the `if` block), and without entering the condition. Additionally, look for missing `else` statements. Exercise edge testing.

## Edge testing

Edge testing verifies that each edge is traversed at least once. An edge could be an `if` condition or a loop statement. But what if we have written our loop's exit condition wrong?

The length and the start of an array are not defined in the same way for all programming languages. For example, languages such as C#, Python, Java, PHP, and Ruby have arrays starting at 0. On the other hand, mathematically-heavy languages such as MATLAB or R, some SQL-based languages, such as PostgreSQL, and older languages such as Cobol or Fortran, have arrays starting at 1. The start of the loops could cause problems if someone programs a loop with a different language in mind.

On the other hand, if the loop condition is `<= array.length`, this could be hitting the end of the array, or if we have `< array.length`, we might be missing the last value.

Here is an example of this:

---

### Calculator.java – method to add multiple numbers

```java
public int addMultiple(int[] numbers) {
    int toAdd = 0;
    for(int i = 0; i < numbers.length - 1; i++) {
        System.out.println("numbers[i]");
            toAdd += numbers[i];
    }
    return toAdd;
}
```

## Loop testing

Loop testing exercises the conditions of the loop: skip the loop, run it once, run it twice, and run it the max numbers of times. If the loop is nested within other loops, iterate with outer loops set to a minimum and test each loop with the aforementioned conditions.

> **Tip**
> To practice loop testing, do not forget to check for all the possible cases in the loops. Check with the condition set to the start, one before (skipping the loop), one after, the condition set to the exit condition, one before the exit, one after, and one or two middle conditions. Do this for every loop with the outer loops set to a minimum.

## Branch testing

This is the most common method for checking for good code coverage. It ensures edges are covered at least once and both conditions of each statement. In order to calculate this, flow charts and abstract syntax trees (more on this in the last section) are usually implemented. The calculation and testing of cyclomatic complexity is a way of understanding the basic sets that guarantee the preceding point. Cyclomatic complexity calculates the maximum number of independent paths of a program; it is also a good tool to use for optimization.

> **Note**
>
> Keep the program as simple as possible, to achieve greater testability. Avoid spaghetti code.

The following code shows an example of an unnecessarily complex program:

### Calculator.java – confusing method to add two numbers

```java
//This method is purposedly confusing and long way of doing
//an addition
public int confusingAdd(int number1, int number2) {
    if (number1 == 0){
        return number2;
    }
    if (number2 == 0){
        return number1;
    }
    if (number1 == Integer.MAX_VALUE && number2 > 0) {
        throw new IndexOutOfBoundsException("first number
            too large");
    }
    if (number2 == Integer.MAX_VALUE && number1 > 0) {
        throw new IndexOutOfBoundsException("second
            number too large");
    }
    if (number1 == Integer.MIN_VALUE && number2 < 0) {
        throw new IndexOutOfBoundsException("first number
            toosmall");
    }
    if (number2 == Integer.MIN_VALUE && number1 < 0) {
```

```
        throw new IndexOutOfBoundsException("second
            number too small");
    }
    if (Integer.MAX_VALUE - number1 - number2 < 0) {
        throw new IndexOutOfBoundsException("the addition
            cannot fit in an integer");
    }
    /* we could add more conditions, the idea is that,
    even a simple addition could get very complicated */
    return number1 + number2;
}
```

## Error testing

Error testing exercises all the possible exceptions that the code could raise and verifies they are caught and handled appropriately. Untested exceptions will cause issues for the users of the software that could be difficult to repair or costly. This can also highlight if we need more code to handle the failure scenarios.

> Tip
>
> Check the code for potential exceptions. Verify all are caught and handled.

## Parameter testing

Parameter testing verifies the parameters sent to the function. Try to find edge cases with them.

> Tip
>
> Check the boundary of each parameter and potential nulls.

Therefore, to provide a good unit test that will really bring up any issues and ensure the code is correct, we need to make sure we are covering all of these verifications. Overall, this means that with a simple addition to the preceding code, our test code could become something like this:

## CalculatorTest.java -

```java
// Tests adding two positive numbers
@Test
public void TestAdd() {
```

```java
    Calculator calculator = new Calculator();
    assertEquals(7, calculator.add(3,4));
}
// Tests adding a negative number and a positive one
@Test
public void TestAddNegative() {
    Calculator calculator = new Calculator();
    assertEquals(1, calculator.add(-3,4));
}
// Tests adding two numbers that overflow an integer.
@Test
public void TestAddOverflow() {
    Calculator calculator = new Calculator();
    assertThrows(IndexOutOfBoundsException.class, () ->
        { calculator.add(2147483647,4);});
}
/*Tests adding a number that would just be the top to
  fit an integer number (this is, the case just before
  the overflow).*/
@Test
public void TestAddMax() {
    Calculator calculator = new Calculator();
    assertEquals(calculator.add(0, 2147483647),
                2147483647);
}
// this could also be 1 + 2147483646 and keep 0 test
// separated
/* Tests adding the minimum number that could fit in an
   integer */
@Test
public void TestAddMin() {
    Calculator calculator = new Calculator();
    assertEquals(calculator.add(0, -2147483648),
                -2147483648);
}
/* Tests adding a negative number that would overflow
```

```
        an integer */
    @Test
    public void TestAddOverflowMin() {
        Calculator calculator = new Calculator();
        assertThrows(IndexOutOfBoundsException.class,
            () -> { calculator.add(-2147483648,-4);});
    }
/* Tests adding two 0 */
@Test
public void TestAddZero() {
        Calculator calculator = new Calculator();
        assertEquals(0, calculator.add(0, 0));
}
//...
}
```

If we execute the previous method, we can see that the overflow does not raise any exception for add, nor does it produce the expected result. If we switch it with confusingAdd, we will see the exception being raised. All the other tests pass.

Most developers, at this point, ask: "is this really worth it?" The reasons for asking this are as follows:

- There are more lines of code in the test code than in the feature code

- This code could also be wrong or even add more mistakes

- An experienced developer will already consider this issue when writing the code

- The code could have dependencies that make it difficult to test

Let's take a look at why we should carry out this testing:

- For 1: While the total number of lines would indeed be higher, the challenges in these lines are fewer than the challenges of writing feature code. There is also a process that could be (there are tools and plugins to make this happen) automated.

- For 2: If the code gets too complicated, it should also be tested. It is possible to test the testing code.

- For 3: The biggest benefit I have ever found from writing TDD and unit testing is that it makes you a more experienced developer. After some years, you might not need these "safety wheels," but you should still write it as a sign of respect to other developers because you are still human and can make mistakes. Moreover, this is a reminder to think of all the possibilities that can happen in the code and potentially uncover future issues when adding features or refactoring.

- For 4: The next section will take a look into how mocking could help out with the dependencies so that the code can be tested independently. Having good testability will also provide better quality coding, which will result in fewer errors and easier debugging of issues.

Remember this quote from Seneca, a famous Roman Stoic philosopher: "*It is not because things are difficult that we do not dare, it is because we do not dare that they are difficult.*"

How could we make this process simpler? If you go back to the first chapter, we could also apply the steps for automation (given in the section *Automating* the *automation*) on this code. In the last section of this chapter, we will analyze ways of automating the code, but in this case, there is an even simpler way of handling this: using data providers.

## Data providers

Data providers or data sources are data inputs that could be connected to the tests, providing a greater range of values to test against, without the need to explicitly write each of those tests.

Most unit testing engines provide a way (generally by annotations) to perform the description of and the iteration through the data. However, this could also be done manually by having the data in a file, reading the file, and iterating through the read data. I recommend using the engine's function when possible so the code does not have to deal with file-related errors.

The data files could also provide outputs to compare to or just the inputs:

### CalculatorDataProvider.java

```java
package chapter2;
import java.util.ArrayList;
import java.util.stream.Stream;
import org.junit.jupiter.api.extension.ExtensionContext;
import org.junit.jupiter.params.provider.Arguments;
import org.junit.jupiter.params.provider.ArgumentsProvider;
public class CalculatorDataProvider implements
ArgumentsProvider {
    /**
    * Represents the data which is expected to pass.
    * data[X][0] = first number to add for the Xth entry
    * data[X][1] = second number to add for the Xth entry
    * data[X][2] = result of the add method for the Xth
                   entry
    **/
```

```java
public static Object[][] goodData ()
{
    // 3 + 4 = 7
    Object[][] data = new Object[3][3];
    data[0][0] = 3;
    data[0][1] = 4;
    data[0][2] = 7;
    // 0 + 0 = 0
    data[1][0] = 0;
    data[1][1] = 0;
    data[1][2] = 0;
    // 1 + 2147483646 = 2147483647
    data[2][0] = 1;
    data[2][1] = 2147483646;
    data[2][2] = 2147483647;
    return data;
}
/**
* Represents the data that is expected to throw an
  exception
* data[X][0] = first number to add for the Xth entry
* data[X][1] = second number to add for the Xth entry
* No need for Data[X][3] as we will not expect result
**/
public static Object[][] exceptionData()
{
    // 2147483647 + 2147483647
    Object[][] data = new Object[3][3];
    data[0][0] = 2147483647; // or Integer.MAX_VALUE
    data[0][1] = 2147483647;
    // -2147483648 + -2147483648
    data[1][0] = -2147483648; // or Integer.MIN_VALUE
    data[1][1] = -2147483648;
    // 2 + 2147483646 > 2147483647
    data[2][0] = 2;
    data[2][1] = 2147483646;
    return data;
```

```
        }
```

For JUnit 5, we need a method to provide the preceding arguments. We will get one object or the other depending on the name of the method that calls this one. This is done as follows:

## CalculatorDataProvider.java – cont – provide arguments

```java
        @Override
        public Stream<? extends Arguments> provideArguments
            (ExtensionContext context) throws Exception {
            Arguments[] argList;
            int length = 0;
            if (context.getTestMethod().toString().contains(
                "Exception")) {
                length = exceptionData().length;
                argList = new Arguments[length];
                for (int i= 0; i<length ; i++) {
                    argList[i] =
                        Arguments.of(exceptionData()[i]);
                }
            } else {
                length = goodData().length;
                argList = new Arguments[length];
                for (int i= 0; i<length ; i++) {
                    argList[i] = Arguments.of(goodData()[i]);
                }
            }
            return Stream.of(argList);
        }
    }
```

## CalculatorTestWithProvider.java

```java
    package chapter2;
    import static org.junit.jupiter.api.Assertions.*;
    import org.junit.jupiter.params.ParameterizedTest;
    import org.junit.jupiter.params.provider.ArgumentsSource;
```

```
public class CalculatorTestWithDataProvider {
    @ParameterizedTest
    @ArgumentsSource(CalculatorDataProvider.class)
    public void TestAdd (int first, int second, int result)
    {
        Calculator calculator = new Calculator();
        assertEquals(result, calculator.confusingAdd(first,
            second));
    }

    @ParameterizedTest
    @ArgumentsSource(CalculatorDataProvider.class)
    public void TestExceptionsAdd(int first, int second) {
        Calculator calculator = new Calculator();
        assertThrows(IndexOutOfBoundsException.class, () ->
            { calculator.confusingAdd(first, second);});
    }
}
```

Please note that if you need to output an error message per result, you can also add it to the data object and to the assert. If we add it to the data as follows:

```
data[0][3] = "Basic happy path should work"
```

Then we could have our test use it, as in this example:

```
@ParameterizedTest
@ArgumentsSource(CalculatorDataProvider.class)
public void TestAdd(int first, int second,
                    int result, string errorMessage) {
    Calculator calculator = new Calculator();
    assertEquals((calculator.add(first, second), result,
                    errorMessage);
}
```

Most examples of unit testing are quite simple. In this chapter, for instance, we have seen an example of an add method. However, in the real world, the "units of code" you can find generally depend on libraries or other pieces of code. In the next section, we will look at why we should use mocking and how to make it useful.

# Do not kill the mockingbird – how to level up your unit testing with mocking

The goal of unit testing is to make sure that a particular unit of code works on its own, but if you are using a library, it could be hard to force that library to return a specific value or values to ensure that the code developed using it works on its own. Let's see that with an example. Imagine your code does not have access to the internet, and, therefore, the following piece will break if tested directly:

**request_call.py**

```python
import requests
def invoke_get() -> bool:
        response = requests.get("https://www.packtpub.com")
        if response.status_code == 200:
            return True
    return False
```

> **Tip**
>
> Don't panic about this code; we will explain requests and responses in the next chapter.

To test this unit, we want to check the different responses we could get from the library that we are using. Since we cannot easily force *that* library to give us different responses, the next best thing is to "mock" it. We can make our own library that will send us the information we want whenever we want it and verify that our code is working appropriately.

The following is an example of such a library:

**request_mock.py**

```python
import requests
def get_200() -> Response:
    response = requests.Response()
    response.status_code = 200
    return response
```

Then we can modify the previous class to use the mock if instructed to do so:

## request_call_with_own_mock.py

```python
import requests
from request_mock import get_200
def invoke_get(mock) -> bool:
    url = "https://www.packtpub.com"
    response = requests.get(url)
    if mock:
        response = get_200()
    if response.status_code == 200:
        return True
    return False
```

And the consequent test would look like this:

## test_own_mock.py

```python
from unittest import TestCase
from unittest import main
from request_call_with_own_mock import invoke_get
class TestOwnMock(TestCase):
    def test_true(self):
        self.assertTrue(invoke_get(True))


if __name__ == '__main__':
    main()
```

Note that we would have to add more complications to test statuses different from 200 and even more to test response messages.

You might be asking again: is it worth it? We are basically writing another library, even if this one is less challenging. Firstly, these libraries can be shared across the organization, so if anybody needs a fake HTML library, they should be able to re-use this one.

Secondly, luckily for us, this can also be automated by using mocking libraries that let us define mock objects, methods, and results.

Here is what the preceding test code would look like using a mocking library:

**test_mock.py**

```python
from unittest import TestCase
from unittest import main
from unittest.mock import patch
from request_call import invoke_get
class TestMock(TestCase):
    def test_true(self):
        with patch('request.get') as req:
            req.return_value.status_code = 200
            self.assertTrue(invoke_get())
    def test_false(self):
        with patch('request.get') as req:
            req.return_value.status_code = 300
            self.assertFalse(invoke_get())


if __name__ == '__main__':
    main()
```

Using `unittest.mock` functionality, we can inject values without creating the specific mocked class and modify the main class to return different values. It is also much easier to check for different values, as shown in the code. Mocking can be a powerful technique when used correctly. Here, I am only scratching the surface of mocking, and I am using the word to refer to any type of test double. We could also have dummy objects, test spies, test stubs, and fake objects.

It is also possible to count the number of times a method was called. Is this against TDD?

## How does mocking affect TDD?

TDD is a technique in which you are meant to write the test code before you write the feature code. However, some mock tests check things such as measuring the number of times that a particular method is called. Even checking branches and exceptions might require a knowledge of how the code is implemented, which could reach the "mystical prediction level" if we were to write the test code first. Does that mean that TDD is impossible or worthless to do?

My theory here is that the best benefit of TDD is that you think about testing early and constantly and whether you do it just before or just after writing or modifying any line of code is secondary.

## The dangers of mocking

While mocking is beneficial to test that the unit of code is doing what it should correctly, many times, it might result in the repetition of test code, with only the actual calls being used, rather than mocked libraries. Code coverage tools used only during unit testing will not be able to check whether the code is covered later on, promoting this repetition.

Writing unit tests with mocking and good code coverage is highly advisable, but if writing the mocking is becoming complicated, time-consuming, and abstract, in my opinion, it might not provide more value than that writing the unit tests with the connected parts.

With mocking, as with unit testing, I find most of this code to be very manual. We could also automate part of that code. Let's see how to do this in the next section.

# Secret passage – automation at the base of the pyramid

As discussed in the last two sections, automation can be used to avoid repetitive code, both in unit testing and during mocking. In fact, mocking libraries are partially doing that very same thing. Furthermore, some **integrated development environments** (**IDEs**), such as IntelliJ IDEA, already have integrated plugins that can insert code automatically, and even create unit test classes. **Visual Studio Code** (**VS Code**) has several community-driven extensions that can automate these and other tasks that you are currently doing manually.

The first requirement to write automation over code is to be able to extract pieces of that code and give meaning to those pieces. This is usually done using an **abstract syntax tree** (**AST**) to represent this code:

- With Python, the `ast` library (see *[1]* in the *Further reading* section at the end of the chapter) provides the methods to create one

- In C#, they are referred to as "expression trees" *[2]*

- A good example with Java can be found here *[3]*

In IntelliJ IDEA plugins, there is a similar concept called **program structure interface** (**PSI**) *[4]* that provides its own methods and objects (PSI elements), customizable to the language selected. For VS Code, extensions are created in JavaScript or TypeScript. To identify the text of a document, you could use an already created AST for the language you are automating, use the semantic token provider, or directly parse the `TextLines` of a `TextDocument` object.

> **Note**
> TypeScript is a typed superset of JavaScript that compiles to plain JavaScript. It offers classes, modules, and interfaces to help you build robust components.

In the following example, we identify methods in an open document in VS editor, by searching for lines that contain `public` or `protected` followed by " (". Then we produce test code for such methods and log it:

## UnitTestWriting.ts

```
'use strict';
import * as vscode from 'vscode';


export async function extractMethodsFromDocument() {
    const editor = vscode.window.getActiveTextEditor();
    const document = editor.getTextDocument();
    let code = "";
    let method = "";
    // get name of doc to set name of Test Class
    for (let i = 0; i < document.getLineCount(); i++) {
        const line =
            document.getTextOnLine(i).toLowerCase();
        if((line.includes("public") ||
            line.includes("protected")) &&
            line.includes("(")) {
        let lineSplit = line.split(new RegExp(" *("));
        if(lineSplit.length>0) {
            method=lineSplit[0];
        }
    }
    code += "\n\n@Test\npublic void " + method +
            "Test(){assert.false;}";
    }
    console.log(code); // here we could write a file
}
```

Then we create the logic for our extension to register the command:

## extension.ts

```
// Register command
'use strict';
```

```typescript
import * as vscode from 'vscode';
import {extractMethodsFromDocument} from './UnitTestWriting';

export function activate(context: vscode.ExtensionContext) {
    const command = 'ut.getUnit';
    const commandHandler = () => {
        extractMethodsFromDocument();
    };
    context.subscriptions.push(
        vscode.commands.registerCommand(command,
                                        commandHandler));

}
```

Finally, we create our package definition:

## package.json

```json
{
    "name": "getUnitExample",
    "version": "1.0.0",
    "description": "Get unit tests",
    "publisher": "packtpub",
    "author": "packtpub",
    "repository":
        "https://github.com/PacktPublishing/Testing-Time-
        Machines/tree/main/Chapter2",
    "devDependencies": {
        "vscode": "^1.35.0",
        "@typescript-eslint/eslint-plugin": "^5.19.0",
        "@typescript-eslint/parser": "^5.19.0",
        "eslint": "^8.13.0",
        "typescript": "^4.6.3",
        "vscode-dts": "^0.3.3"
    },
    "engines": {
        "vscode": "^1.34.0"
    },
```

```
"contributes": {
    "commands": [
        {
            "command": "ut.getUnit",
            "title": "Get Unit Test"
        }
    ]
},
"activationEvents": [
    "oncommand:ut.getUnit"
],
"main": "./extension.js",
"scripts": {},
"dependencies": {
    "vsce": "^2.7.0"
}
}
```

Running `vsce package` will package the extension in a `.vsix` file that can be imported to VS Code. (Note that you need to execute `npm install -g vsce` before using it. Therefore you need Node.js installed on your pc, and different versions might need slight modifications.) To import this file, you should open the VS Code extension tab and click on the top right corner of that tab (...), where you should see the option **Install from VSIX...**

Customizing your IDE will make you more efficient, and learning how to create your own plugins or extensions for it will help you develop a level of automation that can help your company increase quality and speed from the early stages of development.

## Summary

In this chapter, we learned what types of tools, techniques, and testing are used and performed at the base of the test pyramid. Developers so frequently use testing tools that they do not realize that they are even doing testing.

We reviewed the differences between coverage and good unit testing and how mocking can help when the code gets complicated and coupled.

Finally, we showed a way of automating the base of the pyramid to make these tests easier and more fun to write.

In the next chapter, we will look at the middle part of the test pyramid, and we will review backend testing, such as integration, APIs, and contact and shadow testing. This section of the pyramid can be managed by developers or test experts, depending on the company. For companies shifting left, it might be more common that developers are in charge of writing these tests, maybe overseen by a test expert. If you are a **quality assurance** (**QA**) tester or **software development engineer in test** (**SDET**) hoping to grow in your career, this could be an interesting chapter to read.

## Further reading

- [1] More about the AST library can be found here: `https://docs.python.org/3/library/ast.html`.

- [2] More about expression trees can be found here: `https://docs.microsoft.com/en-us/dotnet/csharp/programming-guide/concepts/expression-trees/`.

- [3] A good example with Java can be found here: `https://www.eclipse.org/articles/Article-JavaCodeManipulation_AST/`.

- [4] Find out more about PSI at the following link: `https://plugins.jetbrains.com/docs/intellij/psi.html`.

- [5] See *XUnit Test Patterns* by Gerard Meszaros for a deeper dive into XUnit test patterns.

# The Secret Passages of the Test Pyramid – the Middle of the Pyramid

In the previous chapter, we explored the concept of the test pyramid and its base. If you skipped that chapter and decided to read this one directly, we recommend that you go over its introduction first to understand the test pyramid concept. In this chapter, we are going to cover the middle of the test pyramid and check all its dimensions, secret passages, and dangers.

In this chapter, we are going to cover the following main topics:

- What is inside the middle of the test pyramid?
- Differences between integration, API, and contract testing
- The secret passages – shadowing testing
- The other side of the middle of the pyramid – performance testing

## Technical requirements

Some degree of programming skill is recommended to get the best out of this chapter.

Developers might be especially interested in this chapter, as well as quality experts (SDET/QA) who want to leverage their careers or currently work at this level of the pyramid.

This chapter uses examples written in various programming languages for everyone's convenience. We recommend reviewing and working with different languages as a self-growth exercise. We have provided the examples so that it's easier for you to play with them in our GitHub: `https://github.com/PacktPublishing/How-to-Test-a-Time-Machine/tree/main/Chapter03`.

# What is inside the middle of the pyramid?

In the 2D models of the test pyramid, this section is generally reserved for "integration" testing, though it could also be used for service testing, system testing, contract testing, API testing, and so on. We can argue that a 3D test pyramid model is especially needed at this level, where many different test types are considered.

In this book, we are going to consider the middle of the test pyramid as the part that contains most of the backend-related testing of an application. I am also including performance testing as a hidden side of this level of the test pyramid.

As we build applications with different evolutions of the server-client architecture, we put more focus on the cloud. Careful testing becomes important here to make sure everything goes smoothly and to quickly and easily identify where exactly the problems are.

I remember once encountering a problem that was caused by a server in a bad state. This problem was extremely difficult to reproduce as the system would direct the requests from the client depending on the load of the servers and other factors. Therefore, it is important to have the system alerting us not only about the issue but also providing enough information in the alert for us to be able to understand why is there an issue and how to fix it.

Applications are interlaced with each other, with different developers in each part that could move to another area anytime within the company or even to a different company. Furthermore, companies are frequently using third-party solutions so that they can focus on their side of the business and save some effort on building unnecessary parts. The communication between each of the services or parts of the application, from within the company or outside of it, needs to be as clear as possible.

We will explore more about CI/CD and the role of testing in these systems in *Chapter 6, Continuous Testing –CI/CD and Other DevOps Concepts You Should Know*. There is a section where you can learn more about cloud testing in *Chapter 9, Having Your Head up in the Clouds*. These two concepts are different dimensions to take into account in the middle of the test pyramid but for simplification, we will not include them in this chapter.

Regarding secret passages at this level, we could consider shadow testing and performance testing. We will explore these later in this chapter.

Note that database testing was mentioned in the previous chapter, although this is part of backend testing because it is more "structural" than "communicative."

Let's look at some definitions of the basic tests that we can find inside the test pyramid before deep diving into their differences.

## Integration testing

These tests are done to verify the behavior between (at least) two components of the system.

> **Note**
>
> In this book, we will consider end-to-end testing, system testing, and service testing as types of integration testing. However, as we will see in the next section, they do not cover all of the integration tests that could be done in an application.

We can also have integrated unit tests. These tests are done to verify the behavior of a unit of code but use connections with other systems rather than being isolated and using mocks. These tests are very important too, but they are commonly covered on many books and other resources.. Therefore, in this chapter, we will focus on the other types of integration tests that might not be as straightforward.

## Service testing

To test that a service is working properly, we generally want to make sure that the points that are exposed to that service are working fine. This is generally done by exposing an API or sending messages, which are tested with the next two points.

## API testing

An **application programming interface** (**API**) is one of the exposed points of a service.

For example, the user's view (what is in a browser or user interface) communicates with some software component (which could be in a remote server or within the user's computer), and this component will do the necessary operations for the application to function.

Every API call requires an immediate response. When we test the API, we want to make sure that each request produces the right responses. For example, we can have a request asking for the available rooms at a hotel and another one for the booked rooms. We can then verify that the sum of both results is equal to the total number of rooms in the hotel.

An API is also a great tool for performing some tests or helping us with our automation. We will explore some of this in the next chapter.

## Contract testing

This is an effective way of testing a service by testing its connections to other services. In the case of contract testing, some information is sent from one service (generally, in one server) to another in a different format, generally by messages that are then consumed by the receiving server and acted upon. There is no expectation for an immediate response.

> **Note**
>
> You could have contracts for APIs, defining the structure of the calls. However, contract testing might be done in services that do not have an API too.

## System testing

System testing involves testing the built system, with all the components integrated. It may also refer to testing the built system in different environments/hardware.

> **Note**
> In this book, we will refer to system testing as the end-to-end testing that is done without considering the frontend.

Some of these definitions could be interpreted in different ways. For example, some people might find that all of them are testing a system that is integrated. Integration and system testing would be considered, therefore, synonyms. For your convenience and clarity, we have highlighted the considerations that we are going to have across this book, which might differ from yours, in the definitions.

In the following sections, we will explore what we mean by the different sections and go into the details and tricks to leverage each the most.

## Understanding API, contract, and integration testing

To further clarify the difference between these tests and identify potential dangers and gaps when testing the middle of the pyramid, we have researched for many years and come up with the best schema that will express this:

Figure 3.1: The bicycle example

Yes! This is a bicycle.

The years of research part was a joke. However, one of the first questions that used to be asked in every interview that had anything to do with testing is "how would you test …" followed by a day-to-day item (generally a pen or similar). Let's put that question to work here since a bicycle is a perfect visual example that could help us find the difference between these tests:

- The wheels are components. We can test that they have the right amount of air, friction, and so on (component testing).

- We can also test the unit of the code that would control the wheels – for example, the radius (unit testing). The frame of the bicycle is another unit, with its own set of tests.

- We could have perfect wheels and a perfect frame, but when we put them both together (integration testing), it does not work. This could be because the connections are wrong (service testing), or it could be that the frame is too short and the wheels are too big, so they hit each other and cannot function.

This is still not end-to-end testing as we are yet to connect all the other components, but the expectations between each of the connected components must be met.

For those of you who are more pragmatics, here is another visual example that covers the difference between the types of testing:

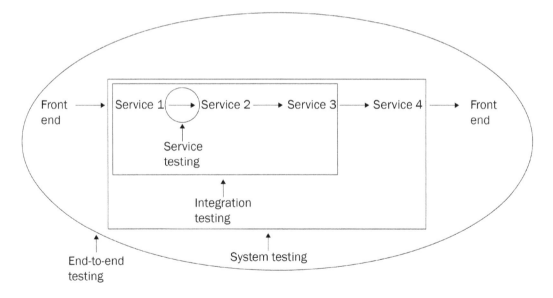

Figure 3.2: Differences between types of testing

Let's explore at each of them in more depth.

## Testing APIs

The most common pattern that websites use to retrieve data from the servers to the client's browser is API calls over the network.

If you are new to APIs and curious about what they look like for a web application, you can just check the **Network** tab on the developer's tool on your browser. There, you will see there are many calls happening in the background when you try to reach a website:

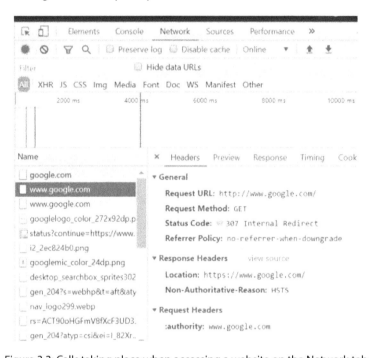

Figure 3.3: Calls taking place when accessing a website on the Network tab

If you click on one of the calls on the left-hand side, you will get some information on the right-hand side. The request URL is the address that the call was trying to hit.

For web calls, some of the possible methods are:

- GET: To request some information from the server. This method is safe (as it does not change anything in the server) and idempotent (identical requests always return identical results until another request changes the values).

- POST: To send some information to the server to create a resource. This method is not idempotent because two calls will produce two resources.

- PUT: Updates an existing resource entirely. This method should be idempotent.

- `DELETE`: Deletes resources. This method should be idempotent.

- `PATCH`: Updates a resource partially.

- `HEAD`: Requests information from the server but without the body. This is thread-safe and idempotent.

- `OPTIONS`: Returns data describing what other methods and operations the server supports.

We will have to verify what methods the API can handle. We can do this by checking the documentation or the API toolset that is being used for development. In the worst-case scenario, which is where we do not have any information, we could send requests to the different methods and try to get information from the responses. This could also be good to validate that the requests fail appropriately.

The next field is also very important. This is the message that we get from the server when the call is executed. In this case, it is 307 (a redirection). Other messages could be on the ranges of:

- **1xx – informational**: 100, 101, 102, 103

- **2xx – success**: 200, 201, 202, 203, 204, 205, 206, 207, 208, 226

- **3xx – redirection**: 300, 301, 302, 303, 304, 305, 306, 307, 308

- **4xx – client errors**: 400, 401, 402, 403, 404, 405, 406, 407, 408, 409, 410, 411, 412, 413, 414, 415, 416, 417, 421, 422, 423, 424, 425, 426, 428, 429, 431

- **5xx – server errors**: 500, 501, 502, 503, 504, 505, 506, 507, 508, 510, 511

We highly recommend you check other error codes online as there are many compilations with funny pictures (do not miss 418!).

Two protocols for sending information between devices you should be familiar with are:

- **Simple Object Access Protocol (SOAP)** *[1]*: Sends information in XML format

- **Representational State Transfer (REST)** *[2]*: Sends information in several formats such as JSON, HTML, XML, and plain text

A nice story *[3]* about them can be found in the *Further reading* section. You can find a good article *[4]* comparing them in the *Further reading* section as well.

Many websites and applications are built using APIs that could be available for direct use. An example of this is Testrail or Jira. See *[5]* and *[6]* in the *Further reading* section for more details. If you want to learn more about APIs, and you are using these sorts of programs, learning to use their APIs could help you build tools for automation and data management that could help you with your career.

Some APIs have easy entry points that do not require authentication (for example, a request to `http://numbersapi.com/42?json` will give you interesting details about the number 42).

Let's look at an example of how to do a general API call with Python:

## non_functional_api_call_template.py

```
import requests
# make a get call
response = requests.get("URL")
# do something with the result
response.status_code # this gives you the status code as
mentioned above
response.json() # this gives you a json with the response on it
# make a post call
response = requests.post("URL", {"json: data"})
```

To give a more precise example, let's see what would happen if we were to call the simple API for numbers:

## get_number_info_call.py

```
import requests
response = requests.get("http://numbersapi.com/42?json")
print(response.status_code)
print(response.json())
```

The result is as follows:

```
200 {'text': '42 is the result given by the web search engines
Google, Wolfram Alpha and Bing when the query "the answer to
life the universe and everything" is entered as a search.',
'number': 42, 'found': True, 'type': 'trivia'}
```

You could use the status code to validate the response and that the request has worked correctly. In this case, the response is simple, but we could still validate some of the fields returned in the JSON, such as if it was found or display just the text in a friendlier way.

Let's look at an example of API testing for this API:

## get_number_info_test.py

```
import unittest
import requests
class GetNumberInfoTest(unittest.TestCase):
```

```
    def test42correct(self):
        response =
            requests.get("http://numbersapi.com/42?json")
        status_code = response.status_code
        self.assertGreaterEqual(status_code, 200)
        self.assertLess(status_code, 400)
    def test_letter_incorrect(self):
        response =
            requests.get("http://numbersapi.com/a?json")
        status_code = response.status_code
        self.assertGreaterEqual(status_code, 400)
if __name__ == '__main__':
    unittest.main()
```

This is just a basic example, but we could test many things around this API, including response messages or even internationalization.

Now, let's look at an example of the Testrail API.

Testrail provides calls to the request library, so instead, we will use its methods that are wrapping these calls. Refer to *[7]* to see how they are done. For example, you can see `requests.get` in line 71 of the aforementioned `testrail.py` file.

Note that first, you need to set up an API key in their website's settings:

---

**print_test_case_id.py**

```
from testrail import APIClient
client = APIClient('http://yourtestrailURL/')
client.user = 'youruseremail' # These two lines are for
authentication
client.password = 'yourApiKey'
case = client.send_get('get_case/id') # This line makes the
request
print(case)
```

The previous method will return something similar to this:

```
{"assignedto_id": 1,"case_id": 1,"custom_expected":
"..","custom_preconds": "..","custom_steps_separated":
[{"content": "Step 1","expected": "Expected Result
```

```
1"},{"content": "Step 2","expected": "Expected Result
2"}],"estimate": "1m 5s","estimate_forecast": null,"id":
100,"priority_id": 2,"run_id": 1,"status_id": 5,"title":
"Verify line spacing on multi-page document","type_id": 4}
```

Here, we can see information about the test case, including its steps, type, and title. You could play with the JSON data to easily retrieve and validate the priority (`case.get('priority_id')`), check how long ago it was created (`case.get('created_on')`), or extract the information to build customized dashboards.

For more information about the type of calls you can make and the results you can expect, see *[8]*. In both of the previous examples, the returned data is short enough for us to check it easily, but when we need to validate several and longer APIs, it is better to use some pre-written JSON libraries to parse the responses into friendlier objects, rather than write the parsing ourselves.

With that, we have showcased not only how to test APIs but how powerful and useful API calls can be.

## Contract testing

A contract between two services is the expected format in which the information that the services require to communicate could be sent. As such, we could test the format of responses from APIs or other sorts of communication, such as message queues.

### *Where to verify contracts*

Contracts can be written and verified at the entry point of the service (to guarantee the right information is arriving at our service) or at the exit point (to guarantee the right information is served from our service).

### Entry contract testing

For each message, we would need to verify that the data is sent to us in the right way. For example, if we get a date, we will need to verify that the format is correct and that it is not too old or newer than we expect. We could use a schema validator for this, but we might also need to perform additional checks on the data received for specific patterns, values, or boundaries.

### Exit contract testing

For each message, we would need to verify that we are sending it the right way, such as by using an XML schema validation or a JSON validator. We could have the validation before sending the message away to verify they are according to the contract that we have exposed with the receiving service.

If we ever need to change our contract – for example, if we want to update the service to provide additional features – we will need to communicate this so that the receiving service can verify the behavior of the old and new messages according to the version.

## What is a schema?

A schema defines the structure of a message containing entities and attributes. They are usually written in XML or JSON format. The following is an example:

## Schema.json

```json
{
    "description":"Testcase",
    "type": "object",
    "properties": {
      "title":{"type":"string"},
      "description":{"type":"string"},
      "id":{"type":"string"},
      "priority":{"type":"string"},
      "steps": {"type":"array"},
      "expected":{"type":"string"}
    },
    "required": [ "title", "description", "id" ]
}}
```

## What to verify

A schema validator could look like this:

## SchemaValidator.cs

```csharp
using System;
using Newtonsoft.Json.Linq;
using Newtonsoft.Json.Schema;
using System.Collections.Generic;

namespace SchemaValidatorSrc
{
    public class SchemaValidator
    {
        // Retrieving schema from before and object from
```

```
        // the response...
        public bool validate(JObject toValidate,
                             JSchema schema)
        {
            IList<string> messages;
            bool valid = toValidate.IsValid(schema,
                                            out messages);
            foreach (string message in messages)
            {
                Console.WriteLine(message);
            }
            /* Instead of writing in console we may want to
            raise an exception here with the message list*/
            return valid;
        }
    }
}
```

By running the preceding code for an invalid type, we will see the corresponding failures and understand which section of the schema was corrupted or wrong.

> **Tip**
> We can reference schemas within other schemas for better readability and reusability. For that, we must use reserved variables or keywords: $schema (which standard is followed), $id (URI for the schema), and $ref (a reference to the URI of another schema).

When using a schema validator is not enough or not possible, we could verify the messages by checking whether the messages are flowing through their expected path.

### Positive entry contract testing

This is done by sending the right messages and expecting some outcome within a timeframe. We could expect the message to go to the right queue or that the database has been updated with the right information. Alternatively, we could check the log files.

### Negative entry contract testing

This is done by sending wrong messages and verifying they arrive at a "delete" queue, or that the right exceptions are logged.

## How to send a queue message

Sometimes, we want to have some system accumulate messages from one server to another. There are two main ways of achieving this:

- **Using a notification system**: One or more servers will subscribe to a notification system, and the other will publish messages. They will all receive the notification at the same time, and it will be deleted afterward. An example of this system is AWS SNS *[9]*.

- **Using a queue**: A server will push messages into the queue and the other servers will take the messages from it, removing them as they read them. An example of this is AWS SQS *[10]*.

In the open source community, we could use Apache Kafka or RabbitMQ to send queues. For simplicity, the following is an example of sending a simple message to a queue called server1Queue using RabbitMQ. You can learn more about them by using the resource links provided in the Further reading section:

### send_message.py

```
import json
import pika
jsonMessage = json.loads('"{\"name\":\"message\",
\"message\":\"test\"}" ')
connection = pika.BlockingConnection(    pika.
ConnectionParameters(host='localhost'))
channel = connection.channel()
channel.queue_declare(queue='server1Queue')
channel.basic_publish(exchange='', routing_key='server1Key',
body=json.dumps(jsonMessage))
# This requires a string, but I assume we would deal with
# json format
connection.close()
```

> **Note**
> You need to have the RabbitMQ server and pika (pip) installed to run the preceding code. You will also need the server key and queue.

## Integration tests with APIs

They can be done similarly to contract testing but involves sending the message in the entry of one service and verifying the behavior in the exit of a different service rather than in the database or the log files. It can also be done by sending an API request rather than a message, as we've seen in previous examples. Internally, this API will produce some communication with another service and produce some output, which we can check using another API:

**integration_api_test.py**

```
import unittest
import requests
class IntegrationAPITest(unittest.TestCase):
    def when_serviceBDataChanges_serviceAResponseChanges(
        self):
        service_A_original_response =
            requests.get(service_A_URL)
        service_B_data_change_request =
            requests.post(service_B_URL,
                          data = {data_to_post})
        service_A_changed_response =
            requests.get(service_A_URL)
        self.assertNotEquals(service_A_original_response,
                             service_A_changed_response)
        self.assertEquals(service_A_changed_response,
                          some_expected_response)
    # Make sure the api goes back to its original state
    # before the end of this test for test consistency (for
    # example with another post reverting the first one).
    if __name__ == '__main__':
        unittest.main()
```

Note that the elements in **bold** have been generalized for this example.

# The secret passages – shadowing testing

Shadowing or mirroring testing is a type of test that is performed to verify two different versions of a system behave in the same way. This testing is especially useful when inheriting a system that already has little testing, to provide a quick check that everything is still in order.

It is done by recording the inputs of a system, as well as its outputs (careful here regarding what data is recorded, including how and who has access to this data), and then verifying them with the outputs in the new system. These replayed tests act like mirrors or shadows of the behavior of the old system over the new system.

Once the new system has been proven correct, we can switch off the servers to get the new code.

## Shadow testing with an API

When an API is available and if the system allows it, it is better to make API calls to both production and pre-production environments with the same inputs so that the outputs are as close to each other as possible, thus decreasing false positives.

## Shadow testing without an API

When an API is not available, then it is more difficult to perform shadow testing since the messages might come into the environments (for example, staging or pre-production and production) in a different order. We must build a process to store both inputs and outputs for both environments, for example, using database tables to store temporarily this information. Then, we need another process to run every time a new input is stored in the table for the staging environment and check if that input was already stored for the production environment and if there are differences between their outputs.

The main danger here is that we could have old inputs and easily cause false positives, so we need to understand when to clear the database data. On the other hand, if the data is cleared too quickly, then we might not find enough data to test in the next deployment.

Metrics can help us get a better understanding of how the system will behave. I suggest having the data stored alongside its date, and having a backup of this data in a table and another table with the current data to test. If the data to test is too small, get data from the backup table; otherwise, use the data to test. After each deployment, move the fresher data to the backup table and clear the data to test so that it is filled up again. If deployments are too fast, they could be done every $X$ number of deployments or on demand.

## Static versus dynamic shadow testing

We could opt to record the inputs once and repeat the calls in production and the server under test (static shadow testing). If we also store the production outputs, this could be less straining for our production servers. However, we would always be testing with the same data and it could not be a significant portion of the actual production traffic, as the requests from users could differ, depending on many factors.

With dynamic shadow testing, we are recording data at a constant rate. This rate could be once a month, once an hour, and so on. If we have metrics in our system, we could get a better understanding of how frequently we should record the data.

## The dangers of shadow testing

The main advantage of shadow testing is that it gives us a quick idea of how well the system is behaving. However, if a failing input is not recorded for some reason, we could miss a potential issue. If the system is very dynamic and the outputs change a lot, then we might find a lot of false positives. The goal should be to have proper testing afterward, keeping this just for migration verification.

An important danger of shadow testing is that production data is being recorded, so the data has to be carefully and securely managed according to the company's data privacy laws. Alternatively, we could process the calls on the go in both systems, which would require us to have the necessary hardware (servers) to process the increasing number of calls. Lastly, we could only send and compare a reduced and random number of calls. The key is that the tests are not happening too late, can stop the deployment if issues are found, and those issues can be logged without logging sensitive data.

## What shadow testing looks like

Shadow testing involves recording calls (these could be fixed or live calls), but the core of the test is a simple comparison of the results. The following code is a simple example of where we would know the value of the calls:

### ShadowTestAPI.cs

```
using System;
using System.Net.Http;
using System.Collections.Generic;
using Microsoft.VisualStudio.TestTools.UnitTesting;
using System.Threading.Tasks;

namespace ShadowTest {

    [TestClass]
    public class ShadowTestAPI {
        private HttpClient client = new HttpClient();

        public static IEnumerable<object[]> DataProvider()
        {
            yield return new object[] { "42",
                "http://numbersapi.com/42?json" } ;
        }
```

```
        [DataTestMethod]
        [DynamicData(nameof(DataProvider),
                 DynamicDataSourceType.Method)]
    public async Task compareData(String
        expectedOutput, String inputCall) {
        HttpResponseMessage actualResponse =
            await client.GetAsync(inputCall);
        String actualString = await
          actualResponse.Content.ReadAsStringAsync();
         Assert.IsTrue(
            actualString.Contains(expectedOutput),
            "The new API is returning wrong data");
    }
  }
}
```

> **Tip**
>
> Note that even though we are using the `UnitTesting` class here, we are doing an integration test and not a unit test; we expect the system to be connected. Please read the previous chapter for tips about unit testing when the method uses external libraries.

There could be different ways of creating these tests, and as discussed previously, they might not be ideal for all applications. Therefore, we should always strive for earlier testing and good integration test plans and make sure we keep user data safe and performance in a good shape.

# The other side of the middle of the pyramid – performance testing

Going back to our bicycle example, we could check how long it works without breaking or how much weight it can support. Performance testing could potentially be measured at any level of the pyramid, but the most common place is the middle.

## The performance… of what?

When we talk about performance testing, most people think about the performance of the system in general, or maybe about each of the services or servers. However, there are many inputs we can check when we are checking performance, including the performance of the client's browser.

## Performance metrics

To understand the performance of our system, we should measure the performance of the different components and keep track of them.

For example, for web applications, we care that our servers are performing well and we can escalate introducing more servers on the go if we get more requests, but verifying the performance of the client (in this case the browser) is also very important.

For mobile applications, we might want to check the memory, cache, and other attributes specific to the phone that the app could be affecting. If a Bluetooth connection is involved, we may also want to verify the behavior with Bluetooth turned off and on, as with Wi-Fi.

For game applications, GPU performance is highly important. Even though the GPU is not something we can control, we could check control that our system is not utilizing more than necessary or producing leaks on it.

Even the heat of the devices could be something we should take into account.

Lastly, we should also test the servers. We should be able to keep track of the status of each of them, including its CPU, memory, and cache, as well as ensure that the requests are received and sent correctly. If we had servers with corrupted code or that were in a bad shape, it would be very difficult to detect if metrics are not present, and the issues would be hard to reproduce as we can't control when the information will be sent to this particular server.

Metrics can give us alerts about whether something is going wrong in our application, but they can also provide very good information about its usage. We could then use these to provide a better service to our users or, as we saw in the previous section, to have good coverage from our shadow testing.

## "Not-performance" testing

We should also have tests in place to verify when a system is not performing, or not performing as expected. When dealing with real users, it could be hard to identify that a system is not performing at all, as it might be the case that no request was sent there. Therefore, we should have a constant request that is being sent and verified constantly. This constant request might actually be part of some other testing happening in production, although they are generally not recommended.

This can be done with other types of applications as well, but it needs to be done carefully so that it does not consume the resources meant for the users.

## Performance in the UI

In the next chapter, we are going to learn about some tricks we can do to improve the performance of our UI tests, but we can also consider the time that the frontend takes to retrieve the objects in a UI test as a potential performance issue. If the objects take too long to appear, or the website takes too long to load, it might be due to a frontend performance issue. For example, the client might be loading too many CSS or JavaScript files that may not even be needed for the view that it's loading. This will result in the website loading slowly, which might have occurred due to the server performance tests. Many beginners in UI testing might consider an object being randomly retrieved in time as flakiness in the test since it increases the waiting time for the object, but this could be pointing toward a frontend performance issue instead.

> **Tips to keep in mind**
>
> Consider having a reasonable global timeout and using it with your explicit waits (but avoid implicit waiting for that time, as it will slow down your tests).
>
> Consider loading (and verifying that this is done) the right files, depending on the view that is loaded. It is preferable to have several separate files than to load bigger multi-purpose ones.

## Load testing

Load testing checks the amount of load that a system can handle. This can be done by simulating users (capacity testing) or sending different load-heavy requests (volume testing). It can be useful for defining the expectations of the system or understanding when an extra server might be needed.

To avoid heavy loads, make sure there is pagination in applications where a lot of data is being sent to and from the user and that this works correctly.

It is important to have an understanding of the expected number of loads the system might need to handle, including the number of users, before planning for load testing.

## Stress testing

Stress testing involves testing the system's ability to recover from failure and do so in a graceful manner, running it under extreme conditions. This could involve running a system for a long time with a high volume in traffic (soak testing), with limited resources, with other applications fighting for those resources, and so on.

It is important to understand the expected possible traffic spikes (especially the dates of sales) of the system and other systems that might have to run beside ours.

## What should a performance test look like?

You could reuse some previous end-to-end tests to add load for a performance test or create specific ones. Some tools allow you to set up the URI you want to hit and the number of requests. This might look more like a configuration than like a test itself.

> **Tip**
>
> Don't reinvent the wheel, enhance it: remember that you might be able to write plugins for the tools you are using, to tailor to specific problems, or for better reporting styles. See the *Secret passage – automation at the base of the pyramid* section in the previous chapter for more information about how to do so.

## Summary

With that, we have reviewed the faces of the middle part of the test pyramid. We started this chapter by defining the differences between the various backend tests and how they relate to each other. We explored API sending, schema validation, and message sending.

Then, we explored a secret passage, known as shadow testing, and when it would be more suitable to use it. Finally, we explored different types of performance testing.

In the next chapter, we will cover the top part of the test pyramid, as well as provide some tricks that you can use to increase the quality of your testing. We will do so by looking at some fun projects you can add to your portfolio.Both quality experts and developers might find the next chapter interesting.

## Further reading

To learn more about the topics that were covered in this chapter, take a look at the following resources:

1. Learn more about SOAP by reading this article: `https://learn.microsoft.com/en-us/archive/msdn-magazine/2000/march/a-young-person-s-guide-to-the-simple-object-access-protocol-soap-increases-interoperability-across-platforms-and-languages`.

2. REST dissertation from Roy Thomas Fielding: `https://www.ics.uci.edu/~fielding/pubs/dissertation/top.htm`.

3. A nice history of SOAP and REST can be found here: `https://thehistoryoftheweb.com/soap-rest-odds/`.

4. A good article comparing SOAP and REST: `https://www.redhat.com/en/topics/integration/whats-the-difference-between-soap-rest`.

5. Testrail API documentation: `https://www.gurock.com/testrail/docs/api`.

6. JIRA information about rest-apis can be found at `https://developer.atlassian.com/server/jira/platform/rest-apis/`. Refer to `https://developer.atlassian.com/server/jira` for more information about JIRA.

7. The particular file that we need for our Python code to use the Testrail API can be downloaded from `https://github.com/gurock/testrail-api/blob/master/python/3.x/testrail.py`.

8.  The API reference for Testrail can be found here: `https://www.gurock.com/testrail/docs/api/reference/tests#userguide-apireference`.

9.  More about Amazon's AWS SNS: `https://aws.amazon.com/sns/`.

10. More about Amazon's AWS SQS: `https://aws.amazon.com/sqs/`.

11. More about Kafka: `https://kafka.apache.org/`.

12. More about RabbitMQ, including more examples than the ones provided in this chapter, can be found at `https://www.rabbitmq.com/`.

# 4
# The Secret Passages of the Test Pyramid – the Top of the Pyramid

In the last two chapters, we explored the concept of the test pyramid, including its base and middle parts. If you have skipped those chapters and went straight to reading this one we recommend that you read at least the introduction to the test pyramid first (covered in *Chapter 2, The Secret Passages of the Test Pyramid – The Base of the Pyramid*) to understand the concept of a three-dimensional test pyramid and the references provided. In this chapter, we are going to cover the top of the test pyramid and check all its dimensions, secret passages, and dangers.

In this chapter, we are going to cover the following main topics:

- What is inside the top of the pyramid?
- The secret passages – headless testing
- The secret passages – making your UI tests more efficient with API calls
- The secret passages – remote execution
- The search for other tests

This chapter will provide an overview of the top of the pyramid from a different perspective than usual. Further tips and patterns will be discussed in the second part of this book.

## Technical requirements

Some degree of programming skill is recommended to get the best out of this chapter.

Quality experts (SDET/QA) might be especially interested in this chapter, as well as developers in a company that has shifted left and who want to leverage their skills with end-to-end testing.

This chapter uses examples written in various programming languages for everyone's convenience. I recommend reviewing and working with different languages as a self-growth exercise. We are also using examples with Cypress and Selenium, but there could be other tools that could be of better use for your system.

The code related to these examples can be found in this book's GitHub repository: `https://github.com/PacktPublishing/How-to-Test-a-Time-Machine/tree/main/Chapter04`.

## What is inside the top of the pyramid?

The top level of the test pyramid consists of those tests that are closest to the user – that is, the tests that are performed in an application (generally) before it is launched to the user. In the 2D models of the test pyramid, this section is generally for end-to-end (frontend automation) and manual testing.

For many years, the top framework for frontend automation was Selenium (with other frameworks based on it), with Appium for mobile apps. Recently, there has been a tendency of looking for faster end-to-end testing, with increasing numbers of headless tests and visual tests to the detriment of pure DOM automation tests. Whichever framework you decide on, you must dedicate the appropriate time to each layer and test so that you do not run into test repetition across the pyramid.

### End-to-end tests/user interface tests/frontend automation

The end-to-end testing that is considered at this level of the test pyramid is generally related to the frontend. In the previous chapter, we saw that you could have some end-to-end integration tests where the frontend was not in the picture. However, at this level of the test pyramid, the application should be almost finished, so we are looking to emulate the users' behavior as much as possible to avoid surprises.

Rather than introducing end-to-end testing, which is covered in many books already, in this chapter, we are going to go through some tricks that will make your end-to-end testing better. We will also cover some side projects so that you can practice them.

# Other dimensions

By the end of this chapter, we will know how to find other tests, two of which might also be different dimensions at the top of the test pyramid. Let's take a look.

## *Accessibility testing*

This type of testing is done to make sure that your application can be used by everybody, including people with disabilities. While this could be done at a lower level, we generally require a UI to be present to test its accessibility.

## *Performance testing*

This type of testing can be done by calling APIs and not necessarily from the UI perspective, but you need your system to be as finished as possible to test its performance in a way that would make sense to the users' behaviors. There are several ways of testing performance, such as from the device's UI (a browser's performance, for example) or testing the device's resource usage (CPU, memory, database, network, and so on). Load testing (testing the system under different loads of tasks upon the maximum expected) and stress testing (testing the system under purposely caused stress, with a higher load or during a long period), are generally covered at this time as well.

## *Security testing*

Security testing can be done at more than the last layer of the pyramid, and there are many books entirely dedicated to it. Since I am not an expert on the subject, I will leave that to the experts, but we have included it in this area of the pyramid so that you keep it in mind. A minimum number of tests could be performed at this level, such as validating inputs and forms with SQL injection, cross-site scripting, denial of service attacks, and others. At other levels, make sure no sensitive data or passwords are stored or/and transmitted. We encourage you to put in place a set of minimum risks and test against them, even if they are not part of this book.

# Secret passages

Within end-to-end testing, there are a few secret passages that will be part of the top of the test pyramid:

- Using an API to make your UI tests more efficient
- Headless testing
- Leveraging remote execution for UI tests (and parallel testing)

By the end of this chapter, you should have an idea of the different dimensions of the test pyramid, so we have prepared a little token for you. Feel free to cut out the following figure and place it somewhere where it could remind you that testing is a many-dimensional area:

> **Important Note**
> To get a good reminder of the test pyramid close, please find a cut out (*Figure 4.3*) provided at the end of the chapter

In this section, we have reviewed the different tests that can be done at this level of the test pyramid. In the next section, you will learn how to make your user interface tests more efficient by using API calls. We are assuming that you can build some end-to-end/user interface tests, but feel free to utilize the references for more tips for your preferred language and framework.

# The secret passages – making your UI tests more efficient with API calls

The goal of every test is to be as small and specific as possible. That way, by taking a glance at the test results, you should be able to tell what went wrong. Making debugging easier is indispensable and valuable, especially as applications grow.

When writing tests, the first question to ask is "what am I trying to test?". The second one is "why?". As obvious as it might sound, these questions are imperative to write powerful and simple tests. At times, we end up testing things that do not correspond to our application's behavior or end up writing a long succession of events to reach the small part that we want to test.

To help with the question of "why am I trying to test" and avoid repetition across tests, we can leverage the use of API calls. While using a UI for testing is a great way of simulating the actual behavior of the users, certain actions are not part of the core of the test that we want to do. We recommend that you read the previous chapter before continuing with this section as it explains how to make API calls in detail.

## Covering repetitive parts

One of the most common issues while writing end-to-end testing is that, when we have a login, we have to log in for most of the tests. If the login fails, most of our test cases will fail since they depend on this login. This gives us very little information about the rest of the functionality. This failure could be very common if our system is getting an object from the DOM to perform the login and this object changes because of some UI refactoring, or even worse, we are using keypresses to perform this login.

One way of avoiding this is by allowing the test user to log in through an API, by creating some cookie or session context that the system could read to validate the login. Then, we can allow the user to navigate to the next view, skipping the login screen. However, we have to keep the security of our system in mind and make sure malicious users can't use this to their advantage, maybe allowing it only under the test environment for some usernames or IP ranges.

We have already mentioned testrails's API file *[1]* and how it handles sending requests internally. We can go to the same file (lines 60 – 66) to verify how the authentication is made:

## Testrail.py

```
. . .
60      auth = str(
61          base64.b64encode(bytes('%s:%s' % (self.user,
                                    self.password), 'utf-8')
62                          ),
63                          'ascii'
64                          ).strip()
65      headers = {'Authorization': 'Basic ' + auth}
...
78      headers['Content-Type'] = 'application/json'
79       response = requests.get(url, headers=headers)
. . .
```

It encodes the user and password in base64 (61), converts it into an ASCII string, and takes away the spaces at the beginning and end of the string (64). Then, in line 65, it adds basic authorization to the headers of the requests. Later, it will add the content type header and pass the URL and the headers to the `requests.get` method in line 79.

These are probably not the only ways you could authenticate in a system, but we hope we have showcased that they are ways of handling logins without the need to use the **graphical user interface** (**GUI**).

As with the previous example, there could be other parts in your application that are repetitive and not part of the test. Keep in mind that we could cover them with an API call, skipping many unnecessary and unstable clicks and object finding.

Finally, sometimes, this login procedure is done over to a third party (such as OAuth). It could be appealing for the users to be able to log in with another application, instead of remembering one more password to access our system. It could also be appealing for us so that we do not have to deal with the authentication part and issues in our code. However, at testing time, it is scary and it complicates the user handling and login process, especially when you are using UI frameworks for this purpose. Using APIs could help us achieve this more simply and reliably than through the UI.

In the previous chapter, we saw how to use testrail's API and identify with it. Let's see how we could use the API to perform a login for a Google account:

1.  First, we need to have a Google cloud/developer account. You can find more information on how to do this here: `https://developers.google.com/`.

2.  Then, we need the credentials (ID and client secret), for when we configure the API. These should be located at `https://console.developers.google.com/`. You should be able to download a `client_secret.json` file that you can store in the same folder as the code we will create next. We will also need to set up the redirect URL for this API (for example, `https://localhost:8080/oauthcallback`).

3.  Then, we need some code for the user to generate an access token. We can get this with the following code:

**gettoken.py**

```python
from google_auth_oauthlib.flow import Flow
flow = Flow.from_client_secrets_file(
    'client_secret.json',
    scopes=['openid',
  'https://www.googleapis.com/auth/userinfo.email',
  'https://www.googleapis.com/auth/userinfo.profile'],
    redirect_uri='https://localhost:8080/oauthcallback')
auth_url, state = flow.authorization_url(
    # Enable offline access so that you can refresh an
    # access token withoutre-prompting the user for
    # permission. Recommended for web server apps.
    access_type='offline',
    # Enable incremental authorization. Recommended as
    # a best practice.
    include_granted_scopes='true')
print(auth_url)
code = input('Enter the authorization code for the above
url: ')
code = code.replace("%2F", "/")
token = flow.fetch_token(code=code)

# retrieve access token
print('access token below')
```

```
print(token['access_token'])
print('duration')
print(token['expires_in'])
```

To get this access code:

I.    Copy the URL that appears in the terminal into a browser and log in. This should give you a URL where the `code=`......& parameter appears.

II.   For your convenience, copy this into a text editor.

III.  Copy everything after `code=` to the next & character and paste that into the terminal that is waiting for input. This should generate a token in the form of some text.

4.   Finally, we can use the token that the previous code generates on the header of a request, as shown in the following code:

## useAccessToken.py

```
import requests
access_token = input('Insert access Token:')
headers = {'Authorization': 'Bearer '+ access_token}
response = requests.get('https://www.googleapis.com/
userinfo/v2/me', headers=headers)
print(response.content)
```

You could enter this access token as a parameter for your tests instead of an input, or retrieve it from some external source at runtime.

Keep in mind that access tokens will need refreshing after some time. If you need to do this, you can run a process in the background to update it. We won't look at this in detail because your tests should run faster than the time it takes for the token to expire. However, if you are running them frequently, you might want to add that logic to avoid having to request new codes constantly. The new access tokens can be stored in some secured place that only the test code has access to and it will be constantly refreshed with new ones.

Examples of how to do this and how to parameterize the request include having a variable retrieved from a config file, an environment variable, a secret value, or a value stored in a database and retrieved at runtime. Check what the safest and most appropriate way for you to do this is.

For more information about Google OAuth checks *[2]* and other OAuth APIs *[3]*.

## Playing with test data

Another possible case of having repetitive parts of code is when we want to add or eliminate items from a list to verify the UI behavior in each case.

For example, if a test is intended to check the user process of eliminating an item from a cart, it needs to have an item first. You will also have the opposite test: adding an item to the cart. It's tempting to have one test depending on the other or have a longer test that checks both things. A better way to proceed is by using the API to add/remove items:

### Validate removal test 1 (pseudocode – not intended to run as code directly):

```
Assuming userTest1 logged in
Perform api call to add item to the cart
Validate removing process
Perform api call to remove all items from the cart as part of
cleanup
```

### Validate addition test 2 (pseudocode – not intended to run as code directly):

```
Assuming userTest2 logged in
Perform api call to make sure that the cart does not have any
items
Perform UI item addition to the cart
Perform api call to remove all items from the cart as part of
cleanup
```

As you can see, both pieces of code will end up removing all the items from the cart as part of the cleanup method, just to make sure the tests leave the system as they found it. Users must be handled well so that the tests run in different accounts and are idempotent. We will learn more about this at the end of this chapter.

Please keep in mind that with this example, we are not checking the results returned by the API. This should have its own API tests and should already be tested by now. Here, we are checking the UI behavior when there are some elements or no elements present, such as error messages, the look and feel, and the hidden or active elements associated with it (can click purchase or not).

We could potentially verify that the API results are incorrect at this stage, but that would require us to have more debugging as the test might fail for many reasons (for example, the XPath used to retrieve the element has changed versus the API failing), besides the issues being harder to fix and implying a context switch.

> **Note**
> It is always preferable to have an API call before a test that has dependency across testing.

## Alternatives to APIs

Rather than making API calls that could potentially alter the system, we could inject the data that we expect the request to return, just like we did in *Chapter 2, The Secret Passages of the Test Pyramid – The Base of the Pyramid*, with mocking but done from the API. The benefit of doing it in an injectable way rather than with actual requests is that the system is going to stay idempotent, no matter how many times you execute the test cases.

Cypress *[4]* has a way of stopping and restarting requests, injecting the expected data in between. This is done by the `intercept` command, as explained on the Cypress website *[5]*.

First, let's learn how to change the response of a website (in this case, `github.com`) with Cypress. We will turn the entire HTML document into an image:

### changeURL.js

```
describe('changeURL', function(){
    it('changesCodeForImage', function() {
    cy.intercept('https://github.com/', (req) => {
        req.reply((res) => {
            res.body = '<img src=
                "https://random-d.uk/api/261.jpg"
                alt="TestingTimeMachines">'
        })
    }).as('newURL')
    cy.visit('https://github.com/')
    cy.wait('@newURL').its('response.body')
      .should('include', 'TestingTimeMachines')
    })
})
```

The result can be seen in the following figure:

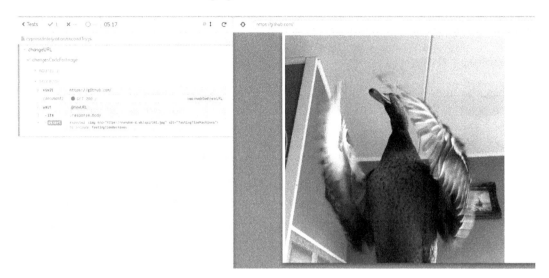

Figure 4.1 – Replacing the github.com source with an image

This is generic code, but we could have a different subset of results. In *Chapter 3*, we inspected a couple of APIs' calls and statuses.

Now, let's see how we would intercept one of the API calls of the previous website with Cypress so that we have a stubbed personalized response instead:

## changingAPI.js

```
describe('changeURL', function(){
    it('changesCodeForImage', function() {
    cy.intercept(
      '/search/count?q=intercept%2F&type=Commits',
      (req) => {
        req.reply((res) => {
            res.body =
              '<span aria-label="Testing time machines"
              title="Changed!" data-view-component="true"
              class="Counter js-codesearch-count
              tooltipped tooltipped-n ml-1 mt-1">Changed!
              </span>                        '
        })
```

```
    }).as('newURL')
  cy.visit('https://github.com/search?q=intercept/')
  cy.wait('@newURL').its('response.body').should(
    'include', 'Changed!')
  })
})
```

The following screenshot shows the results:

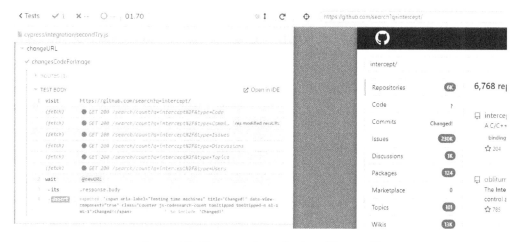

Figure 4.2 – The commit results show "Changed!"

Our test can then check the different UI elements or responses after this has been retrieved. This might not help us with logging in, but it could populate data without the need for us to request it from the actual API (which could be complicated to do at times) or by performing time-consuming and potentially unstable UI actions. Here, we can handle authentication in Cypress:

## authCypress.js:

```
it('Successfully login using headers', function () {
    cy.request({url:
      "https://www.googleapis.com/userinfo/v2/me",
        headers: {
            authorization: 'Bearer access_token'
        }
    }).then((resp) => {
        expect(resp.body).to.have.property('email')
    })
})
```

Here, `access_token` is what we got from `gettoken.py`.

To use the preceding examples, in a terminal, install Cypress:

```
npm install cypress --save-dev
```

Then, save the `js` files in the `integration` folder.

Then, open Cypress and execute the following command:

```
.\node_modules\.bin\cypress open
```

Place your tests under the `cypress/integration` folder to see them in the UI that has been opened and run them. See *[6]* for more information. Injecting data could be done using other frameworks or libraries, such is the case of Selenium. There is a library called **selenium-wire** *[7]* that allows you to do this specifically with Python. However, the Selenium documentation shows how to do that with JavaScript and other languages. This is why it is important that you understand different programming languages and put them to use to your benefit. We are using JavaScript to compare easily with Cypress.

Let's see how we can do an equivalent interception with Selenium:

## changeURLSelenium.js

```javascript
async function changeURLSelenium() {
    const chrome = require('selenium-webdriver/chrome');
    const webdriver = require('selenium-webdriver');
    const http = require(
      'selenium-webdriver/devtools/networkinterceptor');
    const chromedriver = require('chromedriver');
    const assert = require('assert')

    let driver = await new webdriver.Builder()
      .forBrowser('chrome')
      .build();
    const connection =
      await driver.createCDPConnection('page')
    let url = "https://github.com/"
    let httpResponse = new http.HttpResponse(url)
    httpResponse.addHeaders("Content-Type", "UTF-8")
    httpResponse.method = "GET"
```

```
    httpResponse.status = 404
    httpResponse.urlToIntercept = url
    httpResponse.returnBody = "TestingTimeMachines"
    httpResponse.body = "TestingTimeMachines"
    //<img src="https://random-d.uk/api/261.jpg"
    //alt="TestingTimeMachines">'
    await driver.onIntercept(connection, httpResponse,
                             async function () {
      let body = await driver.getPageSource()
      assert.strictEqual(body.includes(
        "TestingTimeMachines"), true,
       `Body contains: ${body}`)
    })
    driver.get(url)
    driver.quit()
}
```

You can run the previous code with Node.js by executing the following command:

```
node changeURLSelenium.js
```

If no error appears, it has passed. Try changing `body.includes` to something other than `"TestingTimeMachines"` to validate this.

For more information on how to do this connection, check *[8]* in the *Further reading* section. By using CDP, we can also attach authentication information directly to the request (using `Network.setExtraHTTPHeaders`). This could help us avoid the UI input of usernames and passwords for basic authentication.

In this section, we reviewed how to make use of an API to facilitate faster execution of important user cases in UI tests. In the next sections, we will learn how to speed up our tests further by using headless browsers and complete difficult tasks, such as testing in different systems, by using remote execution.

## The secret passages – headless testing

To quickly check the code in a browser, we can perform headless testing, which would not include the UI. It performs all the functions needed, but will not show the graphic view, which would result in faster performance and execution of the tests.

The caveat of using such a system is that we are not fully experiencing the same as the users would, so we may miss issues and the ones found would be harder to debug:

- With Cypress, use `./node_modules/.bin/cypress run` instead of `open`, as we did in the preceding examples

- With Selenium, set the options to headless as part of the particular driver of your browser (check the documentation for each)

There are other alternatives to automating with a headless browser; please check if your current framework has one before you decide on another one. Also, if you decide you need a different framework for headless testing, make sure you run benchmarks before arguing which framework performs faster than the other. Also, do not trust other people's benchmarks – each application could be different and they could have set up their system in a specific way you are not aware of.

## The secret passages – remote execution

On occasions, we want to run our tests in a remote browser rather than locally. Sometimes, we want to run them on a particular port of a machine.

This could be useful for many reasons, one of which is that we want to grant access from a particular IP range to a particular port on a particular server. This way, we could avoid setting tokens, headers, or cookies. It could also help us get a more secure and private connection for our tests, and allow us to open and close the port to run the tests only when needed.

Most of the time, we want to test on a specific device or against a device farm. We could own one or it could be from a third party (they will generally provide the syntax and parameters to test remotely on their devices).

When we use tools such as Appium, under the hood we open a server on a specific port on our local computer (by default, `4723`) and by creating a session against it, we are ultimately using a remote browser.

Whichever your reason, keep in mind the benefits of being able to execute the code on a remote browser rather than on a local one.

In these cases, our local client connects with the remote server that is listening to a specific port.

The following is a basic example of how to do this by connecting to our localhost on port `4723`, although there is plenty of documentation about it online:

## ExampleRemoteExecution.py

```
from selenium import webdriver
firefox_options = webdriver.FirefoxOptions()
driver = webdriver.Remote(    command_executor='http://
localhost:8080',    options=firefox_options)driver.get("http://
packtpub.com")
driver.quit()
```

Note that this code won't work unless you have a server on port 8080 of your localhost that will take the URL methods to handle Firefox (in this case) driver methods (such as find_element).

For example, you can use a Selenium server (grid), which can be downloaded from https://www.selenium.dev/downloads/, with a command such as this:

## Starting a Selenium server:

```
java -jar selenium-server-standalone-_x.xx.x.jar -role
webdriver -hub http://localhost:8080/grid/register -port 4444
-browser browserName=firefox
```

Here, you must replace x.xx.x with your Selenium standalone version number, which you downloaded from the preceding link.

Feel free to play around with this feature and take advantage of it as much as possible.

Later in this chapter, in the *Orchestration* subsection, we will learn how to use two machines to communicate during tests.

# The search for other tests

There are many definitions for different types of testing, but the reality is that they are usually mixed in to achieve faster results. The company's budget could end up reducing the types of tests that are being automated. In this section, we will define other types of testing and techniques you can use to identify your particular needs.

## Scaling your tests

Scaling your tests means increasing the number of tests to achieve greater overall quality of the application. This can be done by adding more types of tests or increasing the speed at which the tests are run by executing them across more machines.

## *Vertical scaling*

This is achieved by adding more types of tests (security, accessibility, performance, and so on).

To identify the type of test to be performed, you need to be part of a test architecture design. Depending on your budget, the type of application that you are building, and your users, some tests would take priority. We are not covering security testing in this book since it would likely require its own book and a better expert than I, but that does not mean these tests are less important than those we have covered. Security testing might have a pyramid of its own.

We are also not covering accessibility testing for similar reasons.

As part of your test plan, you could have a test matrix expressing what type of tests you are going to run for each of the services, and if you have them, the numbers of tests too:

|  | Unit tests | API | Contract | Integration | Database | ... |
|---|---|---|---|---|---|---|
| Service 1 | 45 | 20 | 0 | 10 | 15 | |
| Service 2 | ... | | | | | |
| | | | | | | |

Similarly, you can have another one for your endpoints (frontend) if the relationship is not 1:1 and to avoid the table getting too dense:

|  | Unit tests | API | UI | Accessibility | ... | ... |
|---|---|---|---|---|---|---|
| Endpoint 1 | 45 | 30 | 15 | 10 | ... | |
| Endpoint 2 | ... | | | | | |
| | | | | | | |

## Horizontal scaling

Here, you add more machines to your testing. Previously in this chapter, we learned how to achieve horizontality by remotely executing our tests. For example, you could use the cloud to test your application on different systems (more web browsers, mobile, desktop, different OSs, and so on).

Identifying what type of systems will be supported by your application is a design and team decision. If your team does not indicate upfront what systems they are expecting to support, you could take the initiative and provide the systems you will be testing in a support matrix.

As your application evolves, you could add systems to this matrix, and it could be shareable with the users too so that they understand the requirements they will need to use the system.

A support matrix looks like this:

| | Chrome (Windows) | Chrome (Non-Windows) | Firefox | Safari |
|---|---|---|---|---|
| **Supported** | v.43 or higher | v.47 or higher | v.43 or higher | X |
| **With Limitations** | ... | ... | ... | v.7 or higher |
| **Not Supported** | X | ... | ... | ... |

## *Concurrency versus parallelism*

Concurrency and parallelism are two concepts that people pay attention to when they are developing, but they tend to go unnoticed for testing.

### Parallelism

This involves executing several tests at the same time. Normally, this happens on different machines. Sometimes, this is run in an external location, such as in the cloud. We will talk more about the cloud in *Chapter 9, Having Your Head up in the Clouds*.

### Concurrency

This involves executing several tests at the same time, normally, in the same machine. They will be executing partially and out of order.

While it is possible to run different tests on the same machine at the same time if the machine has multiple processors, you can't be certain that it is happening this way. It might seem as though the tests are running at the same time, but the actual execution will be decided by the machine. The processor will be switching tasks so quickly that your eyes will not be able to keep up, but the results of the tests will be randomized.

To use the processor capacity of your machine on demand (and therefore execute the tests in parallel on your machine), you would have to program the tests specifically for this purpose, which could result in complicated and machine-dependent code. At the time of writing, I am not aware of any tools that allow you to do this, and that is for a good reason: it would not be ideal or practical.

Even performance tools that send many requests at the same time have microseconds of difference between them.

However, some applications are built to use the capacity of different processors of a system or machine, so this should be taken into account for testing. Sometimes, issues that come from multiple systems present themselves in the form of random, hard-to-reproduce issues.

## *The importance of independent tests*

It is important to be able to run tests in parallel to achieve more testing in less time. However, some tests reuse users or other test's data. In these cases, it is crucial that each of the tests runs independently and does not alter preconditions for other tests. Tests should provide unique value, but they should also be idempotent – that is, they should provide the same results under the same circumstances every time the test is executed.

We saw an example of this in the first section of this chapter. Imagine two people deciding to run either of our shopping cart tests – for example, the test that removes the item. It could run in this order:

| Person 1 (user 1) | Person 2 (user 1) |
| --- | --- |
| Starts test run | |
| | Starts test run |
| Removes all items from the cart | |
| | Removes all items from the cart |
| Adds an item to the cart | |
| | Adds an item to the cart |
| Ensures the item is in the cart | |
| | Ensures the item is in the cart |
| Removes all items from the cart | |
| | Removes all items from the cart |
| Test passes | Test passes |

However, it could just as easily run in many other ways, like so:

| Person 1 (user 1) | Person 2 (user 1) |
| --- | --- |
| Starts test run | |
| | Starts test run |
| Removes all items from the cart | |
| Adds an item to the cart | |
| | Removes all items from the cart |
| Ensures the item is in the cart | |
| | Adds an item to the cart |
| The test fails (since it is the same user as the person 2 tests removed the item) | Ensures the item is in the cart |

| Removes all items from the cart |                                 |
| ------------------------------- | ------------------------------- |
|                                 | Removes all items from the cart |
|                                 | Test passes                     |

Depending on our luck, both tests might fail, one of them might pass, or both.

If you have no way of identifying when each of the tests is going to be executed (which is the norm for parallel testing), the tests may conflict and fail. These are types of random failures that would add flakiness to your test system and decrease trust in it.

To fix the preceding test, we might want to add unique items to the cart and confirm that a particular item is the one being added. Similarly, we can focus on removing that one item rather than all of the elements.

Item uniqueness can be achieved by adding a timestamp to the item for identification, having a list and picking the items at random (which would not be ideal because it could still produce clashing and flakiness), or by having a system that returns unique items from some database. However, all of these solutions could still fail if too many tests are executed and all of the items are taken. Therefore, we must ensure we will have enough unique items if we go for this solution.

In the same fashion, we could ensure user uniqueness by having a list of users on a database, API, or some other connected system in which we could mark them as being used and release them on test teardown.

Finally, you should pay extra attention to methods to initialize and clean up the tests since they could easily disturb the test's execution. They should be used for common and repeated operations and avoid code repetition, but not to perform tasks that could alter the other tests.

## Orchestration

Understanding how to implement orchestration for your system will make it more efficient. For example, you could orchestrate your test execution to make sure the same test is not executed twice by two different people or on two different devices. Orchestration can also help when two test agents need to communicate, such as when testing a chat functionality. We will see more examples of orchestrating tests in the next chapter.

To orchestrate test data, we could have a system such as the following one:

### orchestrationExample.py

```
from selenium import webdriver
firefox_options = webdriver.FirefoxOptions()
driver1 = webdriver.Remote(command_executor='http://
ip1:4723', options=firefox_options)
```

```
driver1.get("chaturl")
# Assuming login done for user 1, in some way or having
different URLs for different chats
driver2 = webdriver.Remote(command_executor='http://
ip2:4723',   options=firefox_options)
driver2.get("chaturl")
# Assuming login done for user 2
driver1.find_element_by_id("textBox").send_keys("hello user 2")
assertTrue(driver2.find_element_by_id("chatBox").getText().
contains("hello user 2")
driver1.quit()
driver2.quit()
```

The preceding code is just an example and executes everything in the same block. In the next chapter (*Chapter 5, Testing Automation Patterns*), we will learn how to organize the code in a more maintainable and effective way. Also, note that we could parallelize bits of the code if needed by creating threads.

## When to write each T in TDD

We defined **test-driven development** (**TDD**) in *Chapter 2* as "*a technique in which you are meant to write the test code before you write the feature code.*" However, this definition does not clarify what test code you should write before the feature code. As usual, this is up for interpretation, and it depends on each company and application.

Most people will write unit tests before writing the feature code, but for integration tests, they will wait until the system is in place.

Some companies that have test teams separated from the feature code team start with the UI tests at the same time as the feature team starts with the unit test. At times, they wait until the feature code has been started.

Other companies write simple integration tests as unit tests without mocking, and then write the feature code.

Much of the time, the three main tests of the test pyramid will be done in iterations and overlap with the feature code. However, the rest of the tests are usually add-ons and extensions that will be done after the feature is finished or between iterations or updates of the system.

The best procedure is to experiment and do what is best for your team and your system. Do not hesitate to create benchmarks for everything and verify its usefulness but try to do so with an open mind and without any prejudices.

## A word about BDD

**Behavior-driven development** (**BDD**) is a technique for developing software that consists of agreeing on the development of a feature with the entire team and customers before proceeding to develop it. To do so, since BDD is part of TDD, which is where the tests are written first, the tests are described in a common language that can be understood by all parties. An example of this is **Cucumber** To facilitate the transfer of these definitions into code, some tools provide automation over it.

Although we encourage automating coding in this book, we are intentionally skipping BDD techniques and tools on it. This is because they are usually overused and misused. We have seen some test teams being the only ones that are writing and reading these definitions, adding an unnecessary extra layer of complexity to their work instead of reducing it, at times proceeding to write the automation of those test definitions with Selenium.

Lastly, code that is produced with BDD automation is generally not easy to maintain, structure, or reuse, and it is my personal belief that it might not help you grow your development skills. It is OK if you are using BDD for communication and comprehension of features, but be mindful of having side projects if your objective is to get better at programming.

## A word about DDD

**Domain-driven design** (**DDD**) is a series of techniques for developing software in which there is a focus on the core domain, on exploring models in collaboration with the entire team and customers, and on speaking a ubiquitous language within an explicitly bounded context.

The design of a software solution is very important when allocating resources, dividing the work, providing fair deadlines, and achieving a good-quality application. Similar to BDD, the design must be communicated appropriately and must be relatable to the business elements.

DDD, BDD, and TDD can be combined: first, we need a system design that can be understood by everyone. Secondly, we need a way of explaining what each of the features should be doing (and how they should be tested) in a way that can be understood by everyone. Lastly, before any code is created, we should be thinking about how to test it. If you design your system in this way, your chances of having a great-quality app are much higher than otherwise. While there are recommended tools for each, they are not always needed. Therefore, if you are a good developer, you may already be naturally practicing these techniques without even noticing it.

# Summary

When working on a test architecture plan, you must consider every test that could be helpful for your system. Not all systems are the same or have the same needs. In *Chapter 1, Introduction – Finding Your QA Level*, we covered how to discover the needs of our system. In this chapter, we covered some of the main tests the systems should have, as seen from the top of the pyramid to the bottom.

Imagine us pouring water down the pyramid and making sure we cover all the potential gaps throughout the pyramid. We also provided some extra tips and projects you could use to improve your quality at the top of the test pyramid.

In the next part of this book, which covers *Chapter 5*, *Testing Automation Patterns*, to *Chapter 7*, *Mathematics and Algorithms in Testing*, we will provide more useful tips and projects to help you improve your test architecture and career. The next part of this book will build on top of this section. The next chapter will talk about testing automation patterns that will be mostly for end-to-end UI testing. In *Chapter 6*, *Continuous Testing - CI/CD and Other DevOps Concepts You Should Know*, we will talk about CI/CD and testing in DevOps (going back in some way to the middle/bottom of the pyramid). Finally, *Chapter 7*, *Mathematics and Algorithms in Testing*, will cover a subject that might be scary for some: mathematics, and why and how we should learn to love it to create better systems of higher quality.

## Further reading

To learn more about the topics that were covered in this chapter, take a look at the following resources:

- [1] `https://github.com/gurock/testrail-api/blob/master/python/3.x/testrail.py`

- [2] For more information, check `https://docs.cypress.io/api/coapimmands/intercept` `https://github.com/googleapis`, and for more specifically about Google OAuth, check `https://github.com/googleapis/google-api-python-client/blob/main/docs/client-secrets.md`

- [3] You can find more information about how to programmatically enable an OAuth for several different APIs, including Google, at `https://requests-oauthlib.readthedocs.io/`.

- [4] More information about Cypress can be found at `https://docs.cypress.io`

- [5] More information about stabbing a response with Cypress can be found at: `https://requests-oauthlib.readthedocs.io/` and `https://docs.cypress.io/api/coapimmands/intercept#Stubbing-a-response`

- [6] For more information about testing your app with Cypress, go to `https://docs.cypress.io/guides/getting-started/testing-your-app`.

- [7] For more information about selenium-wire, go to `https://pypi.org/project/selenium-wire/` and `https://www.selenium.dev/documentation`

- [8] For more information on how to set up a bidi connection, check out Selenium's documentation: `https://www.selenium.dev/documentation/webdriver/bidirectional/bidi_api/#network-interception` and `https://www.selenium.dev/documentation`

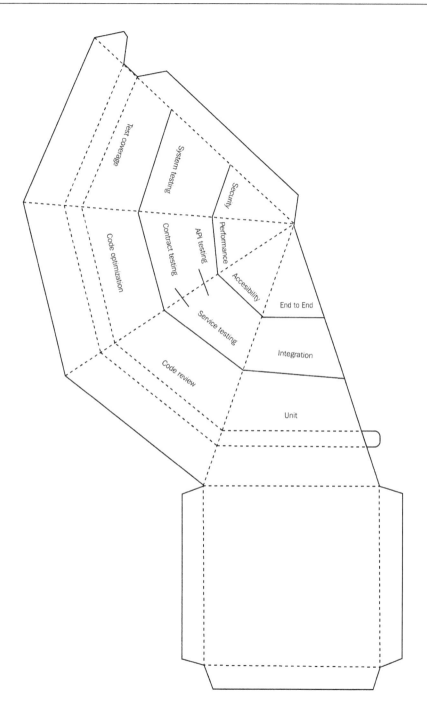

Figure 4.3 – Layered test pyramid

# Part 2
# Changing the Status – Tips for Better Quality

In this part, we will learn about some hands-on tips and tools to achieve better quality and travel to the next level. The topics will be a bit more complex but important to achieve better quality and tools for testing.

This part comprises the following chapters:

- *Chapter 5, Testing Automation Patterns*
- *Chapter 6, Continuous Testing – CI/CD and Other DevOps Concepts You Should Know*
- *Chapter 7, Mathematics and Algorithms in Testing*

# 5
# Testing Automation Patterns

In the last three chapters, we explored the concept of the test pyramid in a deeper and unique way. If you are skipping those chapters and reading this one directly, we recommend you go over the previous one for a better understanding of UI testing.

Most developers are familiar with development design patterns. In testing, we also have patterns, best practices, and ways of dealing with automation writing that have proven useful and can be re-used systematically. The most well-known of these patterns is the **POM**. In this chapter, we are going to cover some versions of it that have been useful in my career, and I hope they help you achieve reliable, maintainable, and faster automation code.

In this chapter, we are going to cover the following main topics:

- The **Page Object Model (POM)** for UI automation
- A **Page Factory Model (PFM)** – an antipattern?
- A **file objects model (FOM)**
- An **enhanced POM (EPOM)**
- A **remote POM (RPOM)**
- Putting it all together
- Automating the automation – a practical example
- Dealing with dynamic objects

## Technical requirements

Some degree of programming skills is recommended to get the best out of this chapter.

Quality experts (SDETs/QA) might be interested in this chapter, as well as developers in a company that has shifted left that want to leverage their skills with end-to-end testing.

This chapter uses examples written in various programming languages. We recommend reviewing and working with different languages as a self-growth exercise; we will provide the implementation for other languages in our GitHub.

Some of these examples are hypothetical, as sometimes it is difficult to find the right kind of system where these sorts of automation would be used (and that can be shared with the readers), but we hope you can find a way to fit these examples to your problem.

> **Note**
>
> In this chapter, we will make references to DOM, according to the *w3org*
>
> The **Document Object Model** (**DOM**) is an **application programming interface** (**API**) for valid HTML and well-formed XML documents. It defines the logical structure of documents and the way a document is accessed and manipulated.

The code related to these examples can be found in this book's GitHub repository: `https://github.com/PacktPublishing/How-to-Test-a-Time-Machine/tree/main/Chapter05`

# The POM for UI automation

The POM is one of the best-known design patterns for writing test frameworks. It consists of separating the model (or test logic) from the pages (the declaration of objects and the locators of elements that those tests interact with). This allows the developer of the test code to find those objects and locators in an effortless way, rather than having to search all the instances of such objects across all the lines of code. The longest the test code, the more tedious this task becomes.

This model creates scalable and maintainable programs that can easily be changed if needed. While this is a good practice for any type of code (and could be considered part of the **don't repeat yourself** (**DRY**) principle), it is especially critical for frontend test code, as the definition of locators and elements tends to change frequently.

> **Note**
>
> **DRY**'s goal is to reduce the redundancy and repetition of software patterns, replacing them with abstraction or using data normalization.

Reusability and better-cleaned code are additional benefits of using the POM.

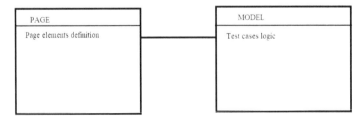

Figure 5.1: Diagram of a POM

In the POM, pages are equivalent to views, which are equivalent to URL endpoints for websites. For other types of applications, such as games, sometimes, it is harder to determine the different screens. In these applications, we can think of pages as forms, windows, screens, and levels. The idea is to group similar types of locators and objects that will be used for grouped tests, too.

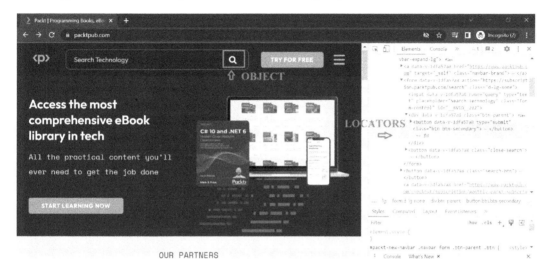

Figure 5.2: Example of a page with a highlighted object and locators

The structure of a POM is as follows.

There is a class per page that defines the objects belonging to the page. In some designs, there is also a way of retrieving them (small methods).

There is a class per logical test group, which defines the test cases and the interactions between the objects. These can map to the pages, but on some occasions, there may be more than one page related to the same test model class. We will see more of this in the **FOM** section.

If we did not use a POM, our code would look something like this:

## NotUsingPOM.java

```
@Test
public void testSearch() {
    //When
    String Url = driver.getCurrentUrl();
    // find search data
    By byIDSearch = By.id("__BVID__336"); // This id
    // might vary a bit, please check is valid when
```

```java
        // running
        WebDriverWait waitForElement = new WebDriverWait(
            this.driver, Duration.ofSeconds(60),
            Duration.ofSeconds(10));
        waitForElement.until(
            ExpectedConditions.presenceOfElementLocated(
                byIDSearch));
        this.driver.findElement(byIDSearch).sendKeys(bookName);
        this.driver.findElement(byIDSearch).sendKeys(
            Keys.ENTER);
        By partialLinkText = By.partialLinkText(bookName);
        // use bookname after book is publishing
        waitForElement.until(
            ExpectedConditions.presenceOfElementLocated(
                partialLinkText));
        // verify url has changed
            Assert.assertNotEquals(Url,
                this.driver.getCurrentUrl());
        }


    @Test
    public void testAddElementToCart() {
        // find search data
        By byIDSearch = By.id("__BVID__336");
        WebDriverWait waitForElement = new WebDriverWait(
            this.driver, Duration.ofSeconds(60),
            Duration.ofSeconds(10));
            waitForElement.until(
        ExpectedConditions.presenceOfElementLocated(
            byIDSearch));
        this.driver.findElement(byIDSearch).sendKeys(bookName);
        this.driver.findElement(byIDSearch).sendKeys(
            Keys.ENTER);
        By partialLinkText = By.partialLinkText(bookName);
        // use bookname after book is publishing
```

```java
        waitForElement.until(
            ExpectedConditions.presenceOfElementLocated(
                partialLinkText));
        this.driver.findElement(partialLinkText).click();
        // ... click add to cart , click cart, verify element
        // is there
    }
    public PageClass(WebDriver driver) {
        this.driver = driver;
        this.waitForElement = new WebDriverWait(this.driver,
         Duration.ofSeconds(60),
         Duration.ofSeconds(10));
    }



    public WebElement getSearchElement() {
        By byIDSearch = By.id("__BVID__336"); // This id might
        // vary a bit, please check is valid when running
        this.waitForElement.until(
            ExpectedConditions.presenceOfElementLocated(
                byIDSearch));
        return this.driver.findElement(byIDSearch);
    }



    public void waitForPartialLink(String bookName) {
        By partialLinkText = By.partialLinkText(bookName);
        this.waitForElement.until(
            ExpectedConditions.presenceOfElementLocated(
                partialLinkText));
    }



    public WebElement getBookPartialLink(String bookName) {
        this.waitForPartialLink(bookName);
        return this.driver.findElement(
```

```
                By.partialLinkText(bookName));
    }
    // Note - we could have a small method to add text to
    // search element and encapsulate this way the actual
    // element
```

Note that we have deliberately omitted the imports and common methods. Please check our GitHub repository to get them.

Using a POM, we will have a page class as follows:

### PageClass.java

```java
package org.packtPub;
import org.openqa.selenium.WebElement;
import org.openqa.selenium.By;
import org.openqa.selenium.WebDriver;
public class PageClass {
private WebDriver driver;
public PageClass(Webdriver driver) {
    this.driver = driver;
}
public WebElement getSearchElement() {
  return driver.findElement(By.id("search"));
}
public WebElement getBookPartialLink(String bookName) {
    return
        driver.findElement(By.partialLinkText(bookName));
}
// Note - we could have a small method to add text to
// search element and encapsulate this way the actual
// element
}
```

And a test class as follows:

## TestCaseClass.java

```java
@Test
public void testSearch() {
    //When
    String Url = driver.getCurrentUrl();
    // find search data
    WebElement searchElement = page.getSearchElement();
    searchElement.sendKeys(bookName);
    searchElement.sendKeys(Keys.ENTER);
    page.waitForPartialLink("achine");
    // verify url has changed
    Assert.assertNotEquals(Url,
        this.driver.getCurrentUrl());
}
@Test
public void testAddElementToCart() {
    page.getSearchElement().sendKeys(bookName);
    page.getSearchElement().sendKeys(Keys.ENTER);
    page.getBookPartialLink("achine").click();
    // ... click add to cart , click cart, verify element
    // is there
}
```

Note that we have omitted the imports and before/after methods; please check our GitHub for the full version. If an object changes the way it is retrieved from any other attributes, the second class does not even need to be recompiled. Only the first class is partially changed.

> **Note**
> From now, and for simplicity, we will only focus on the test search functionality. We have showcased another test method here to prove that in order to change its retrieval, we will have to manually look through the code to find every instance where it was called. There is more repetition, which we avoid using a POM structure.

*Figure 5.1* shows a basic schema for the POM. When dealing with actual applications, things tend to be more complicated than this, with pages sharing components (such as menus), and tests sharing methods.

In this case, we can make use of inheritance to have a better design. In the next section, we will see how to do this.

## Inheritance in the POM

Inheritance is a technique of object-oriented development in which one class (or object) can be based on another class (or object), sharing a similar implementation and forming a hierarchy.

Using inheritance in a POM, we can define all the common methods and objects under the main page or test class. Where there is no need to use inheritance in both the pages and the models, we will display this design with both sections implementing inheritance.

Keep in mind that the relationship between models and pages does not necessarily have to be 1:1.

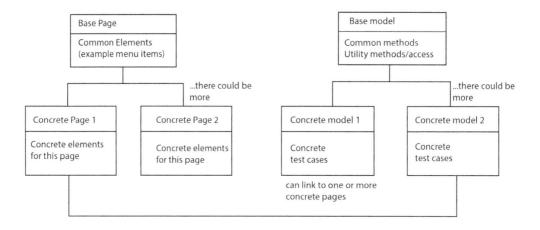

Figure 5.3: Example of inheritance in the POM

Let's see how this would work out if we were to inherit classes to reuse the main menu objects. For simplicity, we are keeping only one model class without inheritance; first, we have our base class, where we have the common class code:

## BasePage.java

```
// omitting imports for readability - check our GitHub repo
// for full code
public class BasePage {
protected WebDriver driver;
```

```
protected WebDriverWait waitForElement;

public  BasePage(WebDriver driver) {
    this.driver = driver;
    this.waitForElement = new WebDriverWait(driver,
        Duration.ofSeconds(60), Duration.ofSeconds(10));
}
public WebElement getSearchElement() {
   By byIDSearch = By.id("__BVID__336"); // This id might
   // vary a bit, please check is valid when running
   this.waitForElement.until(
       ExpectedConditions.presenceOfElementLocated(
           byIDSearch));
   return this.driver.findElement(byIDSearch);
}
// Note - add all the menu elements
}
```

Then, we have one or more page classes that extend (or inherit) the base class with the code that deals with the specific elements of that page:

## MainPage.java

```
// imports omitted for readabilitypublic class MainPage extends
BasePage {
    String freeTrial = "freeTrial";
        String tryForFreeText = "TRY";
    public MainPage(WebDriver webdriver) {
        super(webdriver);
    }
    public WebElement getFreeTrialElement() {
        By partialLink = By.partialLinkText(tryForFreeText);
        this.waitForElement.until(
            ExpectedConditions.presenceOfElementLocated(
                partialLink));
        return driver.findElement(partialLink);
    }
```

```java
    public void waitTrialLoaded() {
        waitForElement.until(
            ExpectedConditions.urlContains(freeTrial));
    }
}
```

Note that we are implementing it this way for continuity, although there are better ways of doing so.

Finally, we have our test class (or classes), which uses the extended classes directly. Classes can access the code of the base page class from the extended classes as if it were written inside of them:

## TestCaseClass.java

```java
// imports removed for readability - please check our
// GitHub repo for the full code
public class TestCaseClass {
    MainPage page;
    WebDriver driver;
    private final String bookName =
        "Testing time machines";

    @BeforeClass
    public void initPages() {
        this.driver = new ChromeDriver();
        this.page = new MainPage(this.driver);
    }

    @BeforeMethod
    public void beforeMethod() {
        this.driver.get("http://www.packtpub.com");
    }

    @AfterClass
    public void afterMethod() {
        this.driver.close();
        this.driver.quit();
    }
```

```
@Test
public void testSearch() {
    String Url = driver.getCurrentUrl();
    // find search data
    WebElement searchElement = page.getSearchElement();
    searchElement.sendKeys(bookName);
    searchElement.sendKeys(Keys.ENTER);
    page.waitForPartialLink(bookName);
    // verify url has changed
    Assert.assertNotEquals(Url,
        this.driver.getCurrentUrl());
}// ...
```

We could add more methods to test the objects on the main page. We could also test the menus in a separate test class; this example is just to showcase that we can access those methods from the model instantiating the main page rather than the base one.

Keep in mind that inheritance might not always be sufficient: sometimes, several model classes will require the use of several page classes as users go back and forth in the system and expect results on the next pages. This is another reason why we keep pages and the model separate – so that they can be used by different models and different models can use different pages without the need to repeat code.

While defining the objects separately is a good practice, it brings up the issue of having to initialize them. In the next section, we will see another pattern, the goal of which is to make this task easier.

# A PFM – an antipattern?

In a PFM, the initialization of all the objects is done in a method called initElements, where the driver that controls the objects is passed as a parameter and the page gets initialized. The benefit of this is that the code for the pages is easier, as we only need the definition of the elements and/or locators (@FindBy) and not any repetitive code for finding those elements (driver.findElement(webElement)), which will be replaced by the use of the webElement object directly.

The page class in this case looks like this:

## PageClass.java

```java
package package com.packtpub.PFM;
import org.openqa.selenium.WebDriver;
import org.openqa.selenium.WebElement;
import org.openqa.selenium.support.FindBy;
import org.openqa.selenium.support.PageFactory;
public class PageClass {
    @FindBy(id=" __BVID__336")
    WebElement searchElement;
    public PageClass(WebDriver driver) {
        PageFactory.initElements(driver, this);
    }
    public WebElement getSearchElement() {
      return searchElement;
    }
    // ...
}
```

This is a promising idea for static pages that are loaded in one go. However, when we deal with dynamic websites, we find that elements are not loaded together, but only under certain conditions. For example, an element might be loaded after the user hovers over another element, opens a menu or a dropdown, alternates through other elements in a carousel fashion, or scrolls down. Using a PFM is considered an antipattern in these cases, as it will produce a lot of false failures and stale element exceptions.

> **Note**
> If writing the code to find the elements is tedious, consider automating it. We will see how to do this at the end of the chapter.

To deal with dynamic pages, it is better to separate the methods for retrieving the objects. Let us see another pattern for doing this.

# Exploring a FOM

In page object model, we could include small methods within pages (such as getters, clicks, and text setters – we could call that method object model). Then we can separate the objects into another document that can be edited without compilation. For example, they could be part of an XML, a CSV, or a JSON file.

The advantage of putting the objects into a separate method that is not part of compilation is obvious: being able to change the elements more quickly and at any time. However, keep in mind that testing over the new objects should be done prior to the production stage to make sure everything works as expected.

Getters and clicks methods could be as short as one line. However, sometimes, these methods can become more complicated than expected and it is interesting to separate them from the test logic. There is also the added benefit that this way, the objects are kept abstracted and encapsulated from the test logic, protecting them from unwanted editions on the model part.

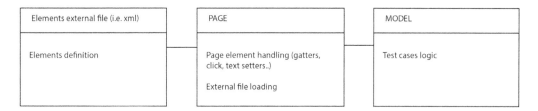

Figure 5.4: Example of FOM design

First, we define our elements in a file, for example, an XML file such as this one:

## Elements.xml

```
<search name="a" type="input" id="__BVID__336" xpath=.../>
<!-- ...-->
```

Note that we could add xpath and other selectors too. The first item in each line is the name that we will use to read it from the class.

Next, we declare the page class that will load those elements:

## FilePageClass.java

```
// imports omitted for readability - check the full code
// in our Github repo
```

```
public class FilePageClass {
    private WebDriver driver;
    private static final Path objectsPath =
        Paths.get(System.getProperty("user.dir"),
        "src", "test", "java", "com", "packtpub", "FOM" ,
        "Elements.xml");
    private Document objDoc;
    public FilePageClass(WebDriver driver) throws
        ParserConfigurationException, SAXException,
        IOException {
        this.driver = driver;
        DocumentBuilderFactory documentBuilderFactory =
            DocumentBuilderFactory.newInstance();
        DocumentBuilder docBuilder =
            documentBuilderFactory.newDocumentBuilder();
        objDoc = docBuilder.parse(objectsPath.toFile());
    /* Letting purposedly the exception flow, as if
        something goes wrong, we want to know before trying
        to find the element */
    }

public WebElement getSearchElement() {
    Element searchElement = (Element)
        objDoc.getElementsByTagName("search").item(0);
    /* we should only get one element, otherwise we are
        doing something wrong. We can add some checkers and
        exceptions to validate this.*/
    By byId = By.id(searchElement.getAttribute("id"));
    return driver.findElement(byId);
}
// Note - we could have a small method to add text to
// search element
}
```

Note that our test class can be the same as in the example for the POM, which highlights the benefit of having a modular design.

Keep in mind that even though this is more code than before, changes to objects can now be made without the need to re-compile the entire class.

Another benefit of this model is that we could have several object locators defined in the XML file and can iterate through them in the page file if they are not found. Since this code starts to get quite repetitive, we can see more about this in the *Automating the automation* section.

An additional benefit of this model is that it allows for easier handling of localization, as we will see in the next section.

## Handling localization

When objects are localized, we want to test whether the strings used are the correct ones. If the UI does not have a lot of easy locators, we might find that our locators are also combined with the locales.

### Handling it from the module

We could also use data providers to handle the locales. This is an easier and cleaner way to proceed, and the advisable approach. This method allows us to have only one method per validation without the need to change the pages (and files).

First, we define our data provider where we add the different data to use for the tests (in this case, the locales that we expect to see):

## LocaleDataProvider.java

```
package com.packtpub.Localization;
import org.testng.annotations.DataProvider;
public class LocaleDataProvider {
    @DataProvider(name = "searchLocale")
    public static Object[][] retrieveSeachLocale() {
        return new Object[][] { { "Search" }, { "busqueda" } };
    }
}
```

Then, our tests will use the providers defined:

## TestCaseClass.java (test)

```
// imports excluded - see full code on our GitHub repo
public class TestCaseClass {
    PageClass page;
```

```java
    private WebDriver driver;
    @BeforeMethod
    public void beforeMethod() {
      this.driver = new ChromeDriver();
      this.driver.get("http://www.packtpub.com");
      this.page = new PageClass(driver);
    }
    @AfterMethod
    public void afterMethod() {
      this.driver.close();
      this.driver.quit();
    }

    @Test(dataProvider="searchLocale",
         dataProviderClass = LocaleDataProvider.class)
    public void testSearchText(String expectedText) {
      WebElement element = page.getSearchElement();
      Assert.assertTrue(element.getAttribute("placeholder")
         .contains(expectedText),
         "expected " + expectedText);
    }// Note - you WILL see this test failing the secont
     // time around, as the website does not have spanish
     // locales...
}
```

### Multiple text strings in the FOM files

We could utilize a FOM design to add the strings needed for the locales in the XML file. For example, we could add attributes (or sub-elements) with those strings. This solution is good if there are only partial changes in the elements (such as text and accessibility tags).

> **Note**
> Only use a FOM to handle locales when there are actual locators changing with the locale. To handle text strings, data providers are preferable.

This time, our elements will include two attributes for the two expected localized strings:

## Elements.xml

```
<search name="a" type="input" id=" __BVID__336 " textEN_
UK="Search" textES_ES="búsqueda" xpath="..."/>
<!--…-->
```

Then, we will load those elements on our page and get these attributes:

## LocalisedFilePageClass.java

```java
// see our github repo for all the imports public class
LocalisedFilePageClass {
    private WebDriver driver;
    private static final Path objectsPath =
        Paths.get(System.getProperty("user.dir"), "src",
        "test", "java", "com", "packtpub",
        "LocalizationFOMText" , "Elements.xml");
    private Document objDoc;

    public LocalisedFilePageClass(WebDriver driver) throws
        Exception {
        this.driver = driver;
        DocumentBuilderFactory documentBuilderFactory =
            DocumentBuilderFactory.newInstance();
        DocumentBuilder docBuilder =
            documentBuilderFactory.newDocumentBuilder();
        objDoc = docBuilder.parse(objectsPath.toFile());
    /* Letting purposedly the exception flow, as if something
    goes wrong, we want to know before trying to find the
    element */
    }

    public WebElement getSearchElement() {
        Element searchElement = (Element)
            objDoc.getElementsByTagName("search").item(0);
```

```
            /* we should only get one element, otherwise we are
            doing something wrong. We can add some checkers and
            exceptions to validate this.*/
            By byId = By.id(searchElement.getAttribute("id"));
            return driver.findElement(byId);
    }


/* ... we could handle this from the module side or add a
method like this to get the different strings from the object:
*/
    public String getSearchElementExpectedText(String
        locale) {
        Element searchElement = (Element)
            objDoc.getElementsByTagName("search").item(0);
        switch(locale) {
            case "ES_ES":
                return
                    searchElement.getAttribute("textES_ES");
            default:
                return
                    searchElement.getAttribute("textEN_UK");
        }
    }


    public String getSearchElementActualText() {
        return
            getSearchElement().getAttribute("placeholder");
    }
}
```

Finally, the test cases will add the pages as before and pass the locale to retrieve the actual string that is in the element for that locale:

## TestCaseClass.java

```
package com.packtpub.LocalizationFOMText;
// see our github repo for all the imports
public class TestCaseClass {
```

```
    LocalisedFilePageClass page;
    private WebDriver driver;

    @BeforeMethod
    public void beforeMethod() throws Exception {
      this.driver = new ChromeDriver();
      this.driver.get("http://www.packtpub.com");
      this.page = new LocalisedFilePageClass(driver);
    }

    @AfterMethod
    public void afterMethod() {
      this.driver.close();
      this.driver.quit();
     }

    @Test
    public void testSearchTextES_ES() {
        Assert.assertTrue(
            page.getSearchElementActualText().contains(
              page.getSearchElementExpectedText("ES_ES")));
    }
    @Test
    public void testSearchTextEN_UK() {
        Assert.assertTrue(
            page.getSearchElementActualText().contains(
              page.getSearchElementExpectedText("EN_UK")));
    }// ...
  }
```

To avoid repeating code, we could use data providers passing the locales as parameters instead of repeating the tests for all locales:

```
LocaleDataProvider.java package
com.packtpub.LocalizationFOMTextWithDataProvider;
import org.testng.annotations.DataProvider;
```

```java
public class LocaleDataProvider {
@DataProvider(name = "localeDataProvider")
    public static Object[][] retrieveData() {
        return new Object[][] { { "EN_UK" }, { "ES_ES" } };
    }
}
```

Then, our test case will only show one test case, but will run one per locale:

## TestCaseClass.java

```java
// ... as before
    @Test(dataProvider="localeDataProvider",
            dataProviderClass= LocaleDataProvider.class)
    public void testSearchText(String locale) {
        Assert.assertTrue(
            page.getSearchElementActualText().contains(
                page.getSearchElementExpectedText(locale)));
    }
// ...
```

### Different files per locale

We could handle this using a FOM with different XML files for each locale. This is the least preferable method, as it has the most repetition and generates files that have to be maintained.

For the first locale, we will create a file with all the elements and all the localized locators and strings:

## ElementsEN_UK.xml

```xml
<search name="a" type="input" id="__BVID__336" text="Search"
xpath="..."/>
<!--...-->
```

We follow the same idea for any other locales:

## ElementsES_ES.xml

```
<search name="a" type="input" id="busqueda" text="búsqueda"
xpath="..."/>
<!--...-->
```

(Note that the tag name stays the same here! id might stay the same or not depending on whether it changes in the development code.)

Then, we can have different documents and filenames:

## LocalisedFilePageClass.java

```java
package com.packtpub.LocalizationFOMMultipleTextFiles;
// check our GitHub repo for imports
public class LocalisedFilePageClass {
    private WebDriver driver;
    private static final Path objectsPath_EN_UK =
        Paths.get(System.getProperty("user.dir"), "src",
                "test", "java", "com", "packtpub",
                "LocalizationFOMMultipleTextFiles" ,
                "ElementsEN_UK.xml");
    private static final Path objectsPath_ES_ES =
        Paths.get(System.getProperty("user.dir"), "src",
                "test", "java", "com", "packtpub",
                "LocalizationFOMMultipleTextFiles" ,
                "ElementsES_ES.xml");
    private Document objDoc;

    public LocalisedFilePageClass(WebDriver driver,
        String locale) throws Exception {
        this.driver = driver;
        DocumentBuilderFactory documentBuilderFactory =
            DocumentBuilderFactory.newInstance();
```

```
        DocumentBuilder docBuilder =
            documentBuilderFactory.newDocumentBuilder();
        switch(locale) {
            case "ES_ES":
                objDoc = docBuilder.parse(
                    objectsPath_ES_ES.toFile());
                break;
            default:
                objDoc = docBuilder.parse(
                    objectsPath_EN_UK.toFile());
                break;
        } /*
           * Letting purposedly the exception flow, as if
             something goes wrong, we want to
           * know before trying to find the element
           */
    }
    public WebElement getSearchElement() {
        Element searchElement = (Element)
            objDoc.getElementsByTagName("search").item(0);
            /* we should only get one element, otherwise we are
               doing something wrong. We can add some checkers
               and exceptions to validate this.*/
        By byId = By.id(searchElement.getAttribute("id"));
        return driver.findElement(byId);
    }
    /* ... we could handle this from the module side or add
       a method like this to get the different strings from
       the object: */
    public String getSearchElementExpectedText() {
        Element searchElement = (Element)
            objDoc.getElementsByTagName("search").item(0);
        return searchElement.getAttribute("text");
    }
```

```java
    public String getSearchElementActualText() {
        return
            getSearchElement().getAttribute("placeholder");
    }
}
```

Then, we need a test case class per locale as well:

## EN_UK_TestCaseClass.java

```java
package com.packtpub.LocalizationFOMMultipleTextFiles;
// check our GitHub repo for the imports
public class EN_UK_TestCaseClass {
    LocalisedFilePageClass page;
    private WebDriver driver;
    @BeforeMethod
    public void beforeMethod() throws Exception {
        this.driver = new ChromeDriver();
        this.driver.get("http://www.packtpub.com");
        this.page = new LocalisedFilePageClass(driver,
                                                "EN_UK");
    }

    @AfterMethod
    public void afterMethod() {
        this.driver.close();
        this.driver.quit();
    }

    @Test
    public void testSearchText() {
        Assert.assertTrue(page.getSearchElementActualText().
contains(page.getSearchElementExpectedText()));
    }// ...
}
```

Alternatively, we could have several localized tests in the same module, have an initializing method per locale, and use data providers to run the tests, reducing the code repetition a bit but also initializing the time in every method, making it slow:

## TestCaseClass.java

```java
package com.packtpub.
LocalizationFOMMultipleTextFilesWithDataProviders;
// find imports in our github repo
 public class TestCaseClass {
    LocalisedFilePageClass page;
    private WebDriver driver;

    @BeforeMethod
    public void beforeMethod() throws Exception {
      this.driver = new ChromeDriver();
      this.driver.get("http://www.packtpub.com");
    }

    @AfterMethod
    public void afterMethod() {
      this.driver.close();
      this.driver.quit();
    }

    @Test(dataProvider="localeDataProvider",
          dataProviderClass= LocaleDataProvider.class)
    public void testSearchText(String locale) throws
      Exception {
      page = new LocalisedFilePageClass(driver, locale);
      Assert.assertTrue(page.getSearchElementActualText()
          .contains(page.getSearchElementExpectedText()));
    }
  }
```

> **Note**
> The data provider is the same as in the previous section.

Finally, we could have different page classes for different locales if for some reason the code changed a lot between them or to reduce the loading time if we used data providers.

We have seen how to handle different objects and strings with a FOM and with data providers. Similarly, in the next section, we will see how to handle other types of objects – in this example, screenshots.

# EPOM

One of the reasons that *record/playback* tools are not popular is that they usually produce tests that are hard to maintain and therefore hard to escalate. In the previous sections, we have showcased the benefits of keeping the code in separate sections and ideally, we would like these tools to do so for us. However, even if the tools were to produce code in such a way, having screenshots and different object locators would revert the system to *not being scalable*.

We can enhance a POM to include screenshots and create automation that runs over the record/playback tools to create maintainable and scalable code, reducing the effort of writing repetitive lines. This will also work for other types of application-related elements:

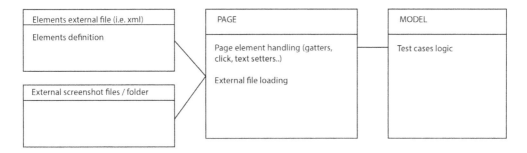

Figure 5.5: Example of design of EPOM

For the following code example, we are going to use Python. The reason for this is that we will be using a library that allows you to find objects using a screenshot (the Airtest project). For more information about this library, you can visit `airtest.netease.com`.

We will define our main page class and add the definition of the screenshots and some code to find the right one as follows (note that the name of your screenshots might differ from these):

## packt_page.py

```
from airtest.core.api import *
import os


auto_setup(__file__)
```

```
class packt_page:
    def __init__(self):
        self.chartElement =
            Template('tpl1631473475812.png',
            record_pos=(0, 0), resolution=(1080, 1920))

        self.chartElementMobile =
            Template(r"tpl1631473465222.png",
            record_pos=(0, 0), resolution= (1080, 1920))
        self.chartElements = [self.chartElement,
                                self.chartElementMobile]
        # define the rest of the elements


    def click_search(self, driver):
        for element in self.chartElements:
            if (driver.assert_template(element)):
                driver.airtest_touch(element)
                break
    # we could and should define more elements or methods here
```

Then, the test will load the page and use the elements, just as we have seen in previous sections:

## packt_test.py

```
 # -*- encoding=utf8 -*-
__author__ = "Noemi"
from selenium import webdriver
from airtest_selenium.proxy import WebChrome
from selenium.webdriver.common.action_chains import
ActionChains

from airtest.core.api import *
import os
path = os.path.dirname(os.path.dirname(os.path.abspath(__
file__)))
path = os.path.join(path, 'packt_pages')
sys.path.insert(0, path)
```

```
from packt_page import *

driver = WebChrome()
driver.get("http://packtpub.com/")

page = packt_page()

page.click_search(driver)
ActionChains(driver).send_keys("Testing time machines").
perform()
```

The structure of the folders and their contents are shown here:

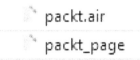

Figure 5.6: Screenshot of the folder structure

> **Note**
>
> The .air extension in the test folder is not really necessary, but good to have if you are using AirtestIDE.

Figure 5.7: Screenshot of the contents of packt.air

Figure 5.8: Screenshot of some of the contents of packt_page

We could have multiple test files in the `.air` folder for the test cases and multiple pages too, although I recommend having separate folders for each so that the screenshots are localized.

If we need to create a new screenshot, we could just use AirtestIDE to retrieve it and copy the text into the actual page or use the code if we feel comfortable doing so.

Another interesting design we could set up would be to organize the screenshots in a folder so that we could iterate through every file in that folder in order to find the right object to select:

## model.py

```
from airtest.core.api import *
import os
# here we import each of the screenshot folders
import os
path = os.path.dirname(os.path.abspath(__file__))
search_screenshot_path = os.path.join(path, 'packt_pages')
# could be elsewhere
files = os.listdir(search_screenshot_path)
class model:
    def __init__(self):
        # define the rest of the elements
        # define the resolutions
        self.resolution1 = (1080, 1920)

    def click_search(self, driver):
        for element in files:
            if (os.path.splitext(element) == ".png"):
                template = Template(
                    element, record_pos=(0, 0),
```

```
                         resolution=self.resolution1)
              if (driver.assert_template(template)):
                  driver.airtest_touch(template)
                  break
   #  ...
```

Another design could be to use a prefix for the names, rather than a folder per object. If there are many objects, this might sound manual, but we could create some standalone automation to rename the files created before or after a certain date or time (5 minutes ago, for example), or files that have a tlp prefix (so they don't have a prefix already) so we can avoid doing so manually. This design would be convenient for the *Putting it all together* section.

In the previous code, we showed how to try to find the elements in different resolutions, which could turn out to be convenient. However, keep in mind that this approach takes away the benefits of having the recording tool to retrieve the code, although we still can use it for screenshots and runs.

### Cleaning the page folder

When dealing with objects such as screenshots, it is important that we clean them after they are no longer needed; otherwise, we end up with a lot of debris. If you decide to use AirtestIDE to use screenshots, there is an option to delete unused images. Keep in mind that if you remove the structure folder, this might not work anymore and you might have to create your own code to identify and remove waste.

If we reduce the manual effort of typing all the templates, then we would need manual effort in either defining the screenshot names so we can remove old ones, or manually removing the ones that are not needed anymore.

The last POM variation that we will look at (RPOM) is needed to handle tests that communicate with each other, such as in the case of a chat.

## RPOM

When the system needs to communicate with two different devices, we need some sort of way of controlling when each of them executes. In this case, we can use an RPOM design such as the following:

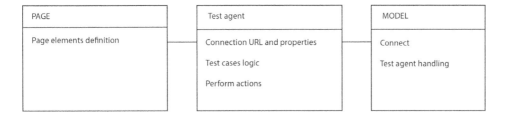

Figure 5.9: Example of an RPOM design

Here, we can have a test agent perform tests as if they were isolated, keeping the logic within them but the decision of the role the agent will play in the test is kept within the model. This is very useful when testing devices that cannot be handled remotely and need a test agent to be installed in them. This is the case for Windows applications, and there is good documentation about agents within Visual Studio's docs: `https://docs.microsoft.com/en-us/visualstudio/test/lab-management/install-configure-test-agents?view=vs-2022`.

Let us see an example of a test agent and a model. The pages are omitted here, as the idea for them is the same as before. The example showcases how to handle a chat within an application (note that we do not need to use Airtest for this section, but we continue to use Python for connectivity):

## testAgent.py

```python
class TestAgent():
    def __init__(self):
        self.chat_page = ChatPage()
        self.timeout = 60
        # init routine, including remote login


    def login(self, user_to):
        return user_to
        #logs in the chat - code omitted


    def test_chat_init(self, user_from, user_to):
        self.login(user_from)
        self.chat_page.wait_for_user_online(user_to,
                                            self.timeout)
        if (not self.chat_page.user_online(user_to)):
            return 1
        self.chat_page.send_chat_to_user(user_to,
                                "Hi from "
                                + user_from)
        self.chat_page.wait_for_chat("Hi from "
                                + user_to,
                                self.timeout)
        if (not self.chat_page.chat_exists("Hi from "
                                + user_to)):
            return 2
```

```
        return 0

    def test_chat_receipt(self, user_from, user_to):
        self.login(user_to)
        self.chat_page.wait_for_chat("Hi from "
                                     + user_from,
                                     self.timeout)
        if (not self.chat_page.chat_exists("Hi from "
                                           + user_from)):
            return 2
            # we associate non existing chat with 2
        self.chat_page.send_chat_to_user(user_from,
                                         "Hi from "
                                         + user_to)
        self.chat_page.wait_for_chat("Hi from "
                                     + user_from,
                                     self.timeout)
        if (not self.chat_page.chat_exists("Hi from "
                                           + user_from)):
            return 2
        return 0
```

Finally, the code from the controller/model would do something like this:

## Controller.py

```
# other imports omitted
import unittest
import subprocess
class Controller(unittest.TestCase):
    def test_chat(self):
        agent1 = {}
        agent1.user = # ...
        agent1.host = # ...
        agent2 = {}
        agent2.user = # ...
        agent2.host = # ...
```

```
        self.run_command("login()", agent1.user,
                        agent1.host)
        # could add check no error here,
        # see next example to understand this method
        self.run_command("login()", agent2.user,
                        agent2.host)
        self.chat_init_response= self.run_command(
            "test_chat_init(\"user1\",\"user2\")",
             agent1.user, agent1.host)
        self.assertEquals(self.chat_init_response, 0)
        # handle the other errors so they are clear on the
        # failure
        self.chat_receipt_response = self.run_command(
            "test_chat_receipt((\"user1\",\"user2\")",
            agent2.user, agent2.host)
        self.assertEquals(self.chat_receipt_response,0)

    def run_command(self, method, user, host):
        command = "ssh " + user + "@" + host+
                "\"path/to/myenv/bin/python\" -c 'from \
                testAgent import *; "+method+"'"
        result = subprocess.Popen(command, shell=True,
            stdout=subprocess.PIPE, stdin=subprocess.PIPE)
        return result.communicate()

    def chat_receipt_response(self):
        return
    # code ommitted
```

In the previous code, we handled the test from the model, as we are remotely connecting with Python, but if we are using another tool such as a Microsoft test agent, then the `assert` methods will be within the agent.

Our design does not always need to be as strict. There are occasions on which our page classes could act as the agents themselves. For example, if we are dealing with mobile automation or several drivers that can be set remotely, we can pass them through to the pages during their initialization without the need to have a separate class. The logic for the test cases would be the model, but the pages would have simple methods to deal with it.

Let us see an example in the following code, in which we will be using `airtest` to connect to two phones:

**mobileChat.py**

```
from airtest.core.api import *
# other imports and setup as in previous airtest example
from poco.drivers.android.uiautomation import
AndroidUiautomationPoco
# class setup omitted too, code could be run without it
android2 = connect_device("Android://127.0.0.1:5037/SERIAL1")
poco2 = AndroidUiautomationPoco(android2)
android1 = connect_device("Android://127.0.0.1:5037/SERIAL2")
poco1 = AndroidUiautomationPoco(android1)
chat_page = chat_page() # you would need to create this!
chat_page.doLogin(poco1) # login omitted from example
chat_page.doLoing(poco2)
chat_page.touch_start_chat_with(poco1, "user2")
chat_page.set_text(poco1, "hi from user1")
assertTrue(chat_page.get_text(poco2, "hi from user1"))
chat_page.set_text(poco2, "hi from user2")
assertTrue(chat_page.get_text(poco1, "hi from user2"))
```

In the previous code, we omitted the page, as we believe the reader can fill in the blanks here but visit our GitHub for examples with more clarity.

We have reviewed different models to handle the automation of the UI. These models are combinable, as we will see in the next section.

## Putting it all together

By now, you should be able to tell which one of the models is best for you. However, keep in mind that all of them (besides the PFM, which we consider an antipattern) are combinable.

Here is how it would look if all the models were working together:

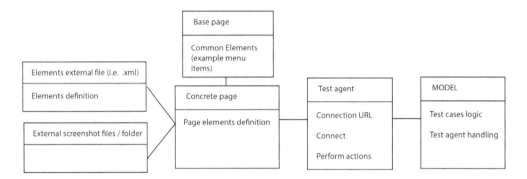

Figure 5.10: Example of models working together

With this, we could have multiple IDs for an object, including screenshots with different resolutions, and the pages would be handled by different agents, which could be called in different servers by the model. This would create powerful, maintainable, and scalable automation.

Whilst having all these models together might not be the right solution for all applications, we have displayed here that we have the flexibility of combining them to best fit our needs. In the next section, we will cover how to automate code repetition in these models so that we can have maintainable, well-designed code that is not tedious to write.

## Automating the automation – a practical example

The general recommendation for test code writing is to make the tests short and simple. This way, we can identify the failing parts easily and clearly. However, sometimes, the time spent creating these simple tests surpasses that of the time of testing the parts that are likely to fail. The key is to find balance and try to reduce the time spent writing repetitive and predictable code.

We have seen before a few examples in which writing automation becomes a tedious and repetitive task: having to type the finders, create methods per object with repetitive header code, and tweak slight changes between them. We also mentioned that if we want the finders to fall through different locators (which will make the tests more robust in case those locators are lost), we will require repetitive code to do this.

Let us break down the process of automation again (as we did in *Automating the automation* section of *Chapter 1, Introduction – Finding Your QA Level*):

1. Recognize automatable tasks.
2. Write some code that does that task for you.
3. Identify when the code needs to be executed.
4. Identify success measures.

We will go point by point and see how these steps can be applied to automation code.

## Recognize automatable tasks

For automation, as we have showcased before, there are repetitive tasks when extracting elements from the DOM and writing the code for those elements' definitions, and searching for other methods for each of the defined elements.

## Write some code that does that task for you

Similarly, to what we did in a FOM, we could *create* an XML file for this, as follows:

### Elements.xml

```
<search type="input" selector="id" id="search"/>
< .../>
```

In order to automate the automation of those elements, we could go through each of the lines of the input (XML) file and create an object with all the elements that are indicated. Then, we can also add a method of the object by its type, as in the following example:

- **For inputs**: A method to enter text
- **For buttons/hrefs**: A method to click
- **For textBoxes**: A method to verify the text

Let us see this in the following example:

### CreateAutomation.java

```java
// imports omitted – check our GitHub repo for full code
public class CreateAutomation {
    private StringBuilder toPrintInPageClass;
    private Document objDoc;
    WebDriver driver = new ChromeDriver();
    public CreateAutomation(String objectsPath, String
        className, String packageName) throws
            ParserConfigurationException,
            SAXException, IOException {
        DocumentBuilderFactory documentBuilderFactory =
            DocumentBuilderFactory.newInstance();
```

```
DocumentBuilder docBuilder =
    documentBuilderFactory.newDocumentBuilder();
toPrintInPageClass.append("package "
    + packageName + ";\n");
toPrintInPageClass.append(
    "import org.openqa.selenium.By;\n");
toPrintInPageClass.append(
    "import org.openqa.selenium.WebElement;\n");
toPrintInPageClass.append(
    "import org.openqa.selenium.WebDriver;\n");
toPrintInPageClass.append("\npublic class "
                            + className + " {");
objDoc = docBuilder.parse(new File(objectsPath));
NodeList nodeList = objDoc.getChildNodes();
for (int i = 0; i < nodeList.getLength(); i++){
    Element o = (Element) nodeList.item(i);
    switch(o.getAttribute("type")) {
        case "input": // define to try by all
            toPrintInPageClass.append("WebElement "
                + o.getAttribute("name") +
                " = driver.findElement(By." +
                o.getAttribute("selector")+
                "(\"" + o.getAttribute("id") +
                  "\"));\n");
            toPrintInPageClass.append("void " +
                o.getAttribute("name") +
                "_type(string toType) { \n\t" +
                 o.getAttribute("name") +
                ".sendKeys(toType); \n}");
            break;
        case "button":
            toPrintInPageClass.append("WebElement "
                + o.getAttribute("name") +
                " = driver.findElement(By." +
                o.getAttribute("selector")+
                "(\"" + o.getAttribute("id") +
```

```
                    "\"));\n");
            toPrintInPageClass.append("void " +
                o.getAttribute("name") + "_click()
                { \n\t" + o.getAttribute("name") +
                ".click(); \n}");
            break;
        case "textBox":
            toPrintInPageClass.append("WebElement "
                + o.getAttribute("name") +
                " = driver.findElement(By." +
                o.getAttribute("selector")+
                "(\"" + o.getAttribute("id") +
                "\"));\n");
            toPrintInPageClass.append("bool "
                + o.getAttribute("name") +
                "_verifyText(string text) {
                \n\treturn " +
                o.getAttribute("name") +
                ".text().equals(text); \n}");
            break;
        }
    }
    BufferedWriter bufferedWriter = new BufferedWriter(
        new FileWriter(new File(className + ".java")));
    bufferedWriter.write(toPrintInPageClass.toString());
    bufferedWriter.flush();
    bufferedWriter.close();
}
```

If instead of using a single locator, we would like to iterate through all the possible locators until the element is found, we could add (parsed to a string) a method such as that in the following example. We are using a map/dictionary data structure to simply map locators and values:

Here's an example of traversing several locators (in Java):

```
public WebElement findElement(Map<String, String>
    locatorValue) {
    WebElement found = null;
```

```
        for (String key : locatorValue.keySet()) {
            found = findElementHelper(key,
                locatorValue.get(key));
            if(found!=null)
                return found;
        }
        return found;
    }


public WebElement findElementHelper(String locator,
                                    String value) {
    switch(locator){
        case "id":
            return driver.findElement(By.id(value));
        case "name":
            return driver.findElement(By.name(value));
    // ...
    }
    return null;
}
```

Spending a little bit of time writing code such as this will reduce the overall time spent writing repetitive code. This code could be used on every page to retrieve all the elements and rebuilt if changes are made to the page, or we could just add more locators to the output. There would also be less human error in the produced code. If there are people on the team without programming experience, they could define the inputs, and learning to execute this code should be straightforward for them to do to update the pages.

### Automating the object retrieval

We have seen before that retrieving the objects from the DOM is still a manual process. We could also automate this part as well.

Note that the structure of the objects we get from a DOM is quite similar to the XML we've seen before. However, in our XML version, the elements did not have any children, so their traversal was easier. When elements have children, siblings, and other types of relations, we can think of them as trees, so if you are thinking of automating a DOM traversal, you should first be familiar with the different traversal algorithms and pick the best one for you. This is beyond the scope of this chapter but we will see more examples of them in *Chapter 7, Mathematics and Algorithms in Testing*.

If we implement all of this, we should not need human intervention to automate the pages, other than running the execution of the code.

## Identifying when the code needs to be executed

If a new page is created with objects from previous pages that have changed a lot, we could execute the entire automation. If the change was minimal, this should be robust enough to resist it, but we could also just change the XML manually. The code will not needed to be compiled.

In addition, we could add into the code functionality a way for modifying an already created class rather than creating a new one each time, but the code needed to do this would be more complicated, and creating a new piece of code to replace the previous automation might be cheaper overall. However, we should make sure that the automatically created objects do not change their designed automated names, as this would require us to change the test case calls to match the new names, and this would be against everything we have seen in this chapter.

A better way of saving updates could be to check whether the input file (XML in our example) has changed elements in the frontend DOM. Since we have also provided a method to retrieve the locators automatically, we could also make the comparison there. This way we will only change the code when needed and keep the automated code more reliable.

Identifying when the page has changed is crucial to identify when to execute automation and can be expensive too. Therefore, we could just run the code execution on the CI/CD pipeline when there are any changes or at a different time preference. We will talk more about CI/CD in the next chapter.

Automating the execution could be another iteration in our automation loop, but in this case, we could tell the computer to run the code at a certain hour, once a week, or once a month. You could add a watcher to the original frontend code and run the code once a change is detected. This could also be part of the CI/CD environment (more on this in *Chapter 6*, Continuous Testing - CI/CD and Other DevOps Concepts You Should Know, up next).

## Identify success measures

After automation is built, it usually needs maintenance and refactoring. The better it is built, the less maintenance you will need.

You should also make sure you have some metrics to ensure your code is correctly automated. This do not need to be automatic, some team feedback during a review meeting might suffice for it.

In this section, we have seen how to automate many simple tests, but you still should be careful not to over-test things, as sometimes it is too expensive or not really needed.

On the other hand, this could be a difficult task to do if we find dynamically created objects or/and dynamically assigned IDs/properties. In the next section, we will see how to deal with some of this.

# Dealing with dynamic object IDs

Although having a good development design with testability in mind is the ideal scenario (and we hope this book helps developers achieve it), the reality is not always as ideal. Sometimes, we inherit old code that has been built as a quick proof of concept or by someone that did not really care for testing so much. In this section, we will see some examples of that and what can we do to automate as much as possible around this. One example is the localized locators that we discussed earlier.

## Items with dynamic locators

Sometimes, it is useful for an application to generate objects automatically. This happens quite frequently in games, where you can find sprites being generated as particles, bullets, or enemies. These may have an auto-generated ID (or another locator).

If that is the case, we should try looking for the object type, using a CSS locator for a particular class or similar.

Here's an example of finding elements using a CSS selector (Java):

```
List<WebElement> bullets = driver.findElements(By.
cssSelector("div# bullets"));
for (WebElement bullet : bullets) {
 // do something
}
```

The do something part could be many things. For example, it could verify a property.

Another way of retrieving elements could depend on the parent object. For example, there may be a table in which rows can be accessed easily if the table has an ID. It could also be similar if this is an HTML list.

Here's an example of finding elements inside a table (in Java):

```
WebElement table = driver.findElement(By.id("bulletTableID"));
List<WebElement> bullets = table.findElements(By.xpath("tr/
td");
for(WebElement bullet: bullets) {
 //    do something
}
```

If this is not an option because we do not have a class or common element to iterate through, autogenerated IDs can be placed instead. Imagine we have a list of users based on input; each of them could have an ID in the order of id1, id2, id3, and so on. However, the user might be present or not on the page we are visiting. If the IDs are autogenerated in this way, we could retrieve all the objects in our DOM that have an ID followed by a number, but we would need to know the maximum number to iterate up to so that we know when to stop the loop. Hopefully, we could retrieve this number by:

- Calling an API

- Creating another user and using the assigned number as top number for iterating, and trying to find the right id iterating backwards from that top number

- Making a guess for user id and keep checking for ids bigger than that guess, if the user id of the guess exists, lower if it does not exist

- Starting from 0 and iterate adding 1, but if the numbers are high up, then it will take you much longer to get to the right one

Here's an example of finding elements using an autogenerated ID selector (in Java):

```java
int maxIterations = 100;
List<WebElement> bullets = driver.findElements(By.
cssSelector("div# bullets"));
int i=0;
while (i<maxIterations){
    WebElement elementI = driver.findElements(By.id("id"
                                                 + i++);
    if (elementI != null) {
    // do something
    }
}
```

> **Note**
> Be careful with objects that might take some time to appear; you might want to retry for a bit. Also, be careful if the `findElements` method throws an exception; you might need `try-catch` instead of having an `if` statement here. This code just exemplifies some cases.

If a series of new unrelated objects is autogenerated, maybe auto-creating the page would be better in that case (as explained in the previous section).

## Angular/React dynamism

Thanks to the introduction of **just-in-time** (**JIT**) compilers, browsers can handle more dynamism in their websites.

> **Note**
>
> A JIT compiler executes computer code that involves compilation during the execution of a program rather than before its execution.

**AngularJS**, **ReactJS**, or **VueJS** are examples of frameworks that allow dynamism. As these frameworks started to become popular, other tools were also created that allow us to deal with this new dynamism. For example, the AngularJS team created **Karma**, which is a **NodeJS** application that allows you to input your tests in the command line and aligns well with **Jasmine** and others tools for testing.

For end-to-end testing, you can check tools such as **Playwright**, **Nightwatch.js**, **Cypress**, and **TestCafe**, among others.

There are many frameworks, extensions, and customizations. Each of them is tailored to a particular case scenario, so if you are building a framework, you need to do good research on them first to understand which one is best suited to you.

Even if you are using one of these tools, you should not forget the design patterns learned in this chapter, because they can really help to maintain clean code and reduce the overall time to write the code.

## Summary

In this chapter, we covered several models for UI automation test code writing, which can help create better-designed frameworks. We also reviewed some ways of automating the repetitive parts of the test code to help you achieve more with less effort.

We have visited different models that could help you differently: with remote topologies, with files with objects, and even with screenshots that can help you with visual automation.

Do you really need all of these testing patterns to write good and stable automation? Well, do you need wardrobes and drawers to be able to find things in your house? The answer is the same.

In the next chapter, we will go back to a lower part of the test pyramid and find out about models that can help us create better testing in CI/CD.

# Continuous Testing – CI/CD and Other DevOps Concepts You Should Know

In the last chapter, we explored different ways of designing **user interface** (**UI**) test automation frameworks. This is found at the top of the pyramid that was introduced in the previous three chapters. The topic of this chapter might seem as though it lands at the bottom of that pyramid, so people working in testing tend to forget about it. However, it is crucial for any system to have the right number of tests that confirm the quality of the code changes that are being implemented and delivered to the user. **Continuous testing** (**CT**) forces the different tests of the pyramid to be run automatically and continuously across all of the deployment phases of the application.

Deciding the approach for this chapter was a tricky task. On the one hand, many concepts could benefit from exact examples. On the other hand, each of the concepts could be written in a different way depending on the tools used. Our advice is to take this chapter as a basic guide and check out the documentation of the tools you would like to learn about in depth.

At the end of the chapter, we will see a basic practical example of how to create continuous testing, even without any **continuous integration/continuous delivery** (**CI/CD**) tools. This could help you strengthen the chapter's concepts and might inspire you to take on some side projects.

In this chapter, we are going to cover the following main topics:

- What is continuous testing?
- CI/CD and other DevOps concepts that you should know
- Types of continuous tests
- Tools for CI/CD
- A basic CT example

# Technical requirements

Some degree of programming skills is recommended to get the best out of this chapter, especially for the *A basic CT example* section. However, it is important that everyone in a company understands the concepts in this chapter.

Therefore, quality experts (SDETs/QA) *should* be interested in this chapter, as well as developers and DevOps team members. Other roles in the company could also benefit from understanding the concepts in this chapter.

This chapter uses partial examples written in pseudocode, as full code would depend on the tools and platforms used for CI. At the end of the chapter, we will see how to automate a `.yml` file using Python. Feel free to try this example with your favorite programming language

The code related to these examples can be found in this book's GitHub repository: `https://github.com/PacktPublishing/How-to-Test-a-Time-Machine/tree/main/Chapter06`

# What is continuous testing?

In order to understand **CT** and how to achieve it, we must first define CI/CD and other core concepts.

## Continuous integration

When several people are working on the same piece of code, issues between the different versions of the code can easily appear. We need a system to keep all the code versions together. The place where code is kept is called a **code repository**.

CI consists of frequently integrating and merging feature code into a shared repository, preferably several times a day. Each integration can then be verified by an automated build and automated tests, which will be part of CT. This is done from within the repository or the different environments set up in the deployment pipeline. The goal is to ensure there are no integration issues and to identify any problems early.

As we evaluate the state of the code after each integration, we can locate and fix integration problems individually. If we do not use this technique, it is harder to debug where the issues came from and determine how to fix them. Therefore, this technique reduces compounding problems.

This technique is also beneficial as problems can be fixed after they have been developed, which increases speed by avoiding time loss in context switching and giving developers confidence in the code being written.

## Continuous delivery

The set of environments where an application is installed before delivering it to users is commonly called a **pipeline**. Sometimes, each of the environments will be related to a version of the code (also known as a branch of the code). We will learn more about this in the *CI/CD and other DevOps concepts you should know* section.

CD is the practice of continuously delivering to the next branch or section of the pipeline, ideally after the tests are executed and have successfully passed. The changes can be pushed frequently and automatically through continuous deployment.

## Continuous deployment

When we finish a version of the code of an application, we need to make sure it is installable and/or works well in different environments.

**CD** sometimes refers to **continuous deployment**, which is the practice of continuously deploying to the environments in order to run tests or to deliver to the user. In this book, we understand continuous deployment as part of CI/CD and will not refer to it as CD in its own right. In other words, the code is integrated (CI), deployed to an environment and tested (CT), and finally delivered to the user (CD).

## Continuous testing

As you can deduce, testing is very important in the CI/CD process. While automated testing is not strictly part of CI, it is typically implied, as we need to detect and locate errors quickly.

Continuous testing has the following goals and benefits :

- It ensures every change is releasable. We can test everything, including deployment at an early and isolated stage.

- It lowers the risk of each release.

- It delivers value more frequently, as reliable deployments mean more releases can be performed in the time it would take to fix the issues.

Many team members would consider DevOps, system reliability engineers, or system administrators to be the only ones responsible for the CI/CD systems, but everybody should be involved and be responsible for it..

As we have seen, continuous testing is very important to ensure all the deployed changes work as expected. Not every test is suitable to run in each of the steps or environments of a pipeline; therefore, it is really important that a test expert is involved in the process of assessing the minimum set of tests that should run and where those tests should be performed. Ideally, we want to keep steps in the pipeline running for less than 10 minutes to provide quick feedback. Otherwise, we would cause the developers to switch contexts. Although this is not always possible, it is good practice. Therefore, test experts should fully understand the deployment process. in order to select the best set of tests at each point.

We have also mentioned before that fixing issues after they are developed helps reduce context switching for developers and makes those issues easier and faster to fix. Therefore, it is useful if the developer who created the code that resulted in the issue is involved in finding and debugging it.

Besides the aforementioned benefits, CI/CD systems increase the involvement of the customer by enabling the possibility of fast and frequent customer feedback on the changes developed.

Now that we understand the general definitions, let us see some other specific concepts that are involved in the CI/CD ecosystem.

# CI/CD and other DevOps concepts that you should know

In this section, we will cover some DevOps concepts you should be aware of and understand in order to be fully involved with and committed to CI/CD systems. Feel free to skip the blocks you are already familiar with.

## Pipelines

A pipeline (also known as a *deployment pipeline*) is the way we organize our deployment system. Each of the steps in our pipeline is matched with an environment that we use to configure the system under development, merge the different versions, and test them.

## Branches

When developing an application, it is convenient to keep different versions of the code, especially when such code is built by different developers at the same time. Each version could correspond with a feature or with a state of the application. The reason that these are called *branches* is from the analogy of a tree, where the main trunk can split into different branches.

A typical design of branches can look as shown in *Figure 6.1*:

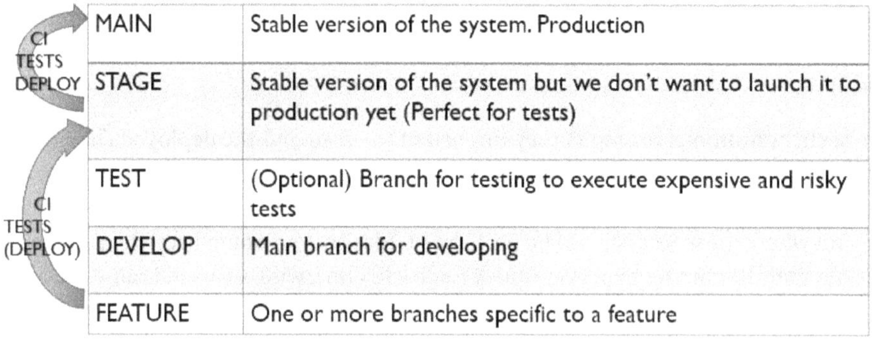

| | | |
|---|---|---|
| CI TESTS DEPLOY | MAIN | Stable version of the system. Production |
| | STAGE | Stable version of the system but we don't want to launch it to production yet (Perfect for tests) |
| CI TESTS (DEPLOY) | TEST | (Optional) Branch for testing to execute expensive and risky tests |
| | DEVELOP | Main branch for developing |
| | FEATURE | One or more branches specific to a feature |

Figure 6.1: A branch system

Each of the steps or stages of a pipeline can correspond to a particular branch, but there can also be different environments for the same branch or even different pipelines for each branch.

The factors that influence the team when deciding how many branches, pipelines, environments, and pipeline steps there should be are many:

- **The number of developers involved**: The more developers, the more merging that could be required of the different features, as there are more branches created. We might also require more pipelines or environments to test the merging of the different features.

- **The complexity of the application**: More complex applications might require more environments to test the deployments.

- **Different systems**: When we have different hardware to install the application on, different environments are needed for testing each of them. We can find the coding of the application broken into different lines of code that are altogether separate from each other, sometimes even being handled and coded by different development teams.

- **The number of teams participating in the development of the application**: Even if the number of developers per team is small, the application's development may be divided into separate teams, each with a different pipeline that needs to be integrated afterward.

- **The number of features being handled at one time**: If the development of the application is incremental, it is common to see features related to branches. However, the application might get divided into different features that each develop subfeatures and all require maintenance and constant development. In this case, we could find different pipelines handling different features, although generally, this would be done by a different development team.

- **Other considerations**: There are other circumstances and business-related issues that would result in different designs of the system.

CI/CD's goal is to automatically execute the CI code after a commit on a branch. Depending on the configuration, it will execute different commands (including tests).

## The difference between a branch and a pipeline step

The way branches are set up can differ a lot between companies and pipelines. It could be the case that a branch matches a pipeline step. However, it is common for code branching to happen before getting the code into the pipeline or for different pipelines altogether.

Pipeline steps are usually related to environments, and they will comprise a set of coding packages of specific versions and a deployment location. Even if our code in particular were to not change between steps, the other set of code packages may change, including the different features of an application which end up uniting to form the final application.

Therefore, we could have features uniting under the same pipeline's step and environment, stage (also called pre-production) environment under another step and production environment under the last step of the same pipeline. Alternatively, we could have features coming from different pipelines and merging to one last pipeline under the same environment, with final stage and production environments as the two final steps..

## Secret variables

It is important that we do not have sensitive data hardcoded into the code base, to protect our company from losses or filtrations. Even when each member of the team is trusted, it is safer that only admins can configure or change the value of certain variables.

When we don't want non-admin contributors to be able to interact with a certain value (such as a password), most CI/CD tools allow the storage and retrieval of sensitive data in what is called **secret variables**.

Admins can set these variables and, in the code, they can be retrieved securely at runtime. Each CI/CD tool has its own libraries to handle these variables. Check your CI/CD tool's documentation for more information on how to work with secret variables.

## Feature toggles/flags

These are used in order to maintain the same version of the code but provide different outputs to the users.

They can be implemented as a condition statement (if a feature on/then). However, having many feature toggles can complicate the code and result in hard-to-fix issues. Therefore, CI/CD tools have libraries to handle them.

Sometimes, branching is fully substituted by feature toggles.

There are many ways of implementing and configuring feature toggles, and it is important we are aware of them when it comes to testing, as we should be testing all the possible versions we could find of the application.

One of the more basic ways of implementing them could be to use a conditional statement, as shown in the following piece of code:

### FeatureToggleExample (pseudocode)

```
if (featureToggle == on) {
// V2 - insert code for this feature toggle here ( for
// example, show in a different colour)
} else {
// V1 - insert feature before the toggle, if it existed
// before (for example, original colour)
}
```

Some CI/CD tools have their own libraries to construct and handle feature toggles, so please refer to their documentation for more information.

Another benefit of using feature toggles is that code merges become easier, as there is less overlapping of parts of code: we add a piece of code, rather than having to change an existing one. The following examples showcase the autogenerated merging issues markings inserted by Git for a full version change versus when there is a feature toggle:

## FullVersion (pseudocode)

```
<<<<<<< HEAD
document.getElementById("featureElement").style.color =
"#ff0000";
// note there could be many other lines here, this is a
// short example
=======
document.getElementById("featureElement").style.color =
"#ff0011";
// note there could be many other different lines here
>>>>>>>.newFeature
```

## Using feature toggle (pseudocode)

```
<<<<<<< HEAD
=======
if (featureEnabled) {
document.getElementById("featureElement").style.color =
"#ff0011";
// note there could be many other lines here, this is a
// short example
} else {
>>>>>>>.newFeature
```

The changes in the HEAD piece of code are not shown as they will stay the same in the new feature, which is much nicer for the merge and easier to review. Most times the merging system will be even smart enough to make these changes automatically without need for human reviewing them at all, as they are mostly additions.

With feature toggles, we can control turning features off or on while deploying everything to the same branch or a different one. They also allow us to quickly turn off a broken feature without the need to redeploy the entire app.

They work very well for some of the deployment strategies and A/B testing, as we will see in the next section.

## Load balancers

The *load* of a system is the number of requests traveling across it. A load balancer is an entity (usually a server) that distributes the traffic around other sets of servers. There are different ways of distributing the traffic; the easiest example is to do so by sending equal amount of traffic across all servers. However, sometimes we want to distribute the traffic according to regions, the different performances of the server, the different applications installed (we could have some servers that take inner requests only), and the version of those applications. The load balancer is what decides which traffic goes where, and should also be able to switch the traffic on demand.

## Rolling

Modern applications are commonly distributed over the cloud and deployed on independent servers. When an application deals with several production servers and we need to deploy code to all of the servers, there are different ways of doing so. **Rolling** refers to deploying an application one server at a time. It is also known as incremental or phased roll-outs. The newly deployed servers generally (although not always) receive user traffic straight away (after some testing).

The advantage of this type of deployment configuration is that we only need one server turned off for user traffic at a time, meaning it is cheaper to deploy this way.

## Re-creating

We have several production servers where we need to deploy an application and the deployments are done on all those servers at once. In order to work with a configuration such as this one, we would need several extra servers to do the switch, but then all the users would be able to see the new application at the same time. This is positive in the sense that an update will be seen by all the users at the same time, but also negative in the sense that potential issues will hit all of our users too.

If the application does not support having multiple versions running together, this is the best way to proceed with deployments.

A subset of the re-creating strategy is known as **big-bang**. In this strategy, all the servers are deployed at once and the old code is replaced with new code, without needing extra servers to do it. This strategy is more dangerous than the others analyzed, and it is only advised for small applications under a restricted budget.

In the next section, we will see how different deployment testing is performed for each of the two previously mentioned deployment systems (rolling and re-creating).

## Rollback

When an issue is found, or something goes wrong, rolling back is the process of leaving everything as it was before the deployment or merge started.

## Metrics, alarms, and reports

**Metrics** help us measure different aspects of our application, such as the state of the servers, performance peaks/issues, and the number of users hitting each server.

If a metric goes above or below a certain value, we can set up **alarms** to notify us of such conditions, so we can fix issues before they affect the users.

**Reports** are snapshots of the metrics at any given point. They can help us review the metrics and we can save them to compare them to other snapshots and make decisions and predictions about our system based on them. We can also use them to find differences between metrics on different branches or with different feature toggles enabled/disabled, including how well they are performing, and whether they are attracting more or fewer users.

The faster we can determine quality, the faster issues are found and fixed, and, therefore, the faster we can deploy an application. Selecting the right type and number of tests is crucial for fast and high-quality deployments. Balance is key.

In the next section, we will see the types of continuous tests and where and when to run them in our pipelines.

# Types of continuous tests

From *Chapter 2, The Secret Passages of the Test Pyramid – The Base of the Pyramid,* to *Chapter 4, The Secret Passages of the Test Pyramid – The Top of the Pyramid*, we explored different types of testing in the test pyramid. In this section, we will see how and where these (and other tests) should be handled through the different steps of our pipeline system, continuously executing them to ensure the high quality of our deployment system.

## Unit tests/code review testing

Continuing the example of *Figure 6.1*, we can commit a feature branch at any time. When we are happy with the contents of the feature branch, ideally after some testing (such as unit tests and code reviews) has taken place, we can merge it to the next branch, including the main trunk.

The test expert should configure the coverage and verify it. We can also create tools to check and ensure the quality of the code being committed before undergoing a code merge.

For more tips about tests and tools for the base of the pyramid, check out *Chapter 2, The Secret Passages of the Test Pyramid – The Base of the Pyramid* .

## Integration tests/system tests/contract tests/API tests

These tests can run in any of the steps of the pipeline to ensure things work as expected before proceeding to the next step. They can be more extensive than the other tests, as they test the feature's complete functionality, but they are run less frequently.

It is possible, although not advisable, to execute longer tests without stopping some steps of the pipeline. This is done in cases where teams are waiting on others to work on their feature, and when other tests are covering the conversing parts. However, there must be some solution in place to stop the pipeline if these long tests fail, so it is recommended that instead you leave the extensive tests to run after all those sequential teams catch up.

## Build verification test

**Build verification tests** (**BVTs**) are a subset of basic tests that ensure that the most important functionality of the application (or feature) works as expected. These are generally integration or **end-to-end** (**E2E**)/UI tests. When the pipeline merges with other pipelines, we should repeat all the BVTs of the different pipelines merged, to make sure the merge has not caused any unexpected issues.

The goal of the BVTs is to provide fast feedback to the developer about the general state of the code at the point of execution and to ensure the most critical parts of the system are still functioning as expected.

If there are any failures in the BVTs, the pipeline should stop before proceeding to the next step and alert the developers so they can fix the issues.

## E2E/UI tests

At the end of the pipelines, before deploying to another pipeline or to the user, we should run all the E2E/UI tests related to the feature under development. These are very similarly placed to integration tests, but for when there is a frontend. However, for simplification, we have not taken them into account in the following diagrams.

The following diagram represents the possible configuration of a pipeline:

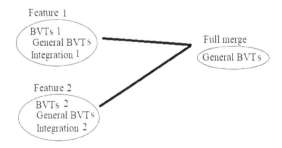

Figure 6.2: Pipeline configuration 1

In the previous configuration, all steps (which are feature steps, and might have several steps each) have extensive testing. However, after the merge, there is only a smaller set of BVTs to ensure that the basic application's functionality is not broken. The basic BVTs are also run in each of the steps to make sure the particular feature or team is not responsible for breaking the app's main functionality. However, if one feature breaks another, it is harder to catch. Let's see a different example of configuration:

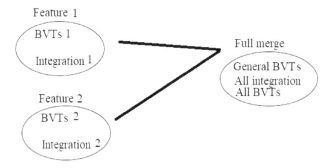

Figure 6.3: Pipeline configuration 2

In this example, the most extensive tests are done at the final merge. In this case, if some of the teams are breaking the main functionality, it's harder to catch what caused the issue. It also takes longer to deploy, as we run more tests, which could potentially be redundant. However, as we re-run everything at the end, it is safer. This can be a good system when the steps (teams/features) are likely to cross over or break one another. Let's see another configuration:

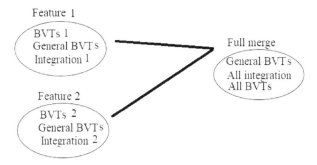

Figure 6.4: Pipeline configuration 3

In this configuration, there are many more tests. If any step breaks the main functionality, it will be caught before merging. The full integration and set of BVTs ensure nobody breaks anything from another feature after a merge. However, we land into more redundancy.

There are many possibilities; we could also have BVTs coming from other features in each of the steps, but that would result in more redundancy. We could have integration running just at the very end with the full merge if integration happens across features.

As usual, testing is a matter of finding the balance between the quality of the deployment speed and the quality of the feature itself, while trying to avoid test redundancy. Perhaps, if features are truly independent, we can just use the first system, and then maybe we just need some extra assurance along the way or at the very end.

Before deciding on a configuration, please also keep in mind that we can run extensive tests on a production server before enabling it for users. See more in the upcoming *Deployment testing* subsection.

In the stage/staging or pre-prod environment, everything should look just as in the production environment. This is a great opportunity to perform extensive testing and ensure all potential tests are caught and resolved before releasing the application to users. We could have a dedicated server to run those tests to make sure nothing is broken after all the merges.

If the tests are extensive, we could run them overnight or in parallel systems to reduce the load.

When we finally merge to production or main, CI will execute new tests and deploy to the production server automatically with the commit. If tests are broken, CI will revert to the state of the stable version.

## Deployment testing

In the following sections, we will cover ways of testing related to different ways of running deployments on servers. It is very important that we have different ways of continuously monitoring the servers for different factors, to make sure no memory or other resources are leaking and damaging the response of the servers. We should also track which server holds which version of the application and which version serves which users/regions.

### Red-black testing

Some of the machines will have the next version of the application and others will not. To differentiate between them, the servers with one version will be called *red*, and the ones with the other version, *black*. However, the machines with the new version will not be distributed to the users until they are all deployed and ready. Then we will switch all users to the other color at once.

Extensive testing for this sort of deployment can be done in all systems at the same time, as they will converse in parallel, and they will take the same amount of time.

### Blue-green testing

Similar to before, the versions are related to colors. However, in this case, both versions can be seen by users. Eventually, all the traffic will switch to green. This is the deployment system used by canary testing or A/B testing, as we will see in the following definitions.

Extensive testing for this sort of deployment can be done in the first instance, while the others could involve less testing (some BVTs) to validate the application and server behave in a stable manner.

> **Note**
> These definitions are interpreted this way in this book as we consider it be a cleaner division, although many large applications would use them indistinctively, so please be mindful of that.

### Canary testing

During canary testing, in a blue-green deployment, small versions of the application are rolled out slowly to a small percentage of users. Once the deployed versions have been proven to perform well, these versions are deployed to more and more users until it becomes the actual version for all of the users.

Our first batch of *canaries* could be servers turned off from any users and we could use them for extensive testing, including all the tests across all the pipelines, as they could run in parallel and the redundancy wouldn't affect us much.

The way we distribute the canaries across the users is generally not conditioned and frequently happens in a stratified way. The application must be able to support old and new versions working together. Ideally, all the users will move to the same version at some point.

### A/B testing

A/B deployments are a type of blue-green deployment similar to canaries, but in which the distribution to the users is determined by specific conditions. An example of such a condition might be a subset of users (beta users, for example), or regions. The goal of A/B testing is to understand and gather information about the user's response/interaction with different versions of the systems.

One could think of a version, the "A" version, and another, the "B" version, but this system does not necessarily work with just two versions. Multiple versions can co-exist even after the testing is done. For example, if we find some version of the app more successful in some regions than others, we might want to keep that version in that region. In those cases, the use of feature toggles is highly recommended to keep the versions stable.

Each version should be tested in its own environment to make sure all the features are working as expected. This might complicate the test cases and the tracking of issues a bit, so the test experts must have an organized way of tracking the several versions and/or toggles and where and how to test them.

## Performance monitoring

When dealing with servers and the cloud, different things could go wrong and we should constantly keep an eye on them to measure them. We should monitor resources, such as the CPU, memory, network, and other details (such as the GPU if the application uses it heavily), to make sure there are no strange peaks, downs, or incremental increases that could cause our server to malfunction or our application to break.

Sometimes, we also want to check the performance of the client, but we will not monitor it as part of the CI/CD pipeline but as a standalone performance test. Those tests could be done as part of the developer's environment, in a middle environment, or in a pre-production environment. We should use an actual production server to test them, as they might influence the production database or other resources. Therefore, we monitor the performance rather than running performance tests in the production environment.

When we consider that we may have a few users interacting with the application at any point, but we want to make sure the application is behaving as it should, or verify the different servers are still up and running, we might want to *fabricate traffic* to run it on the server. This traffic should be marked in some way for us to differentiate it from *real* user traffic, for easier tracking and understanding the potential issues.

In the previous sections, we have seen the different ways of deploying and testing an application continuously. In the next section, we will talk about tools that will help us achieve it.

# Tools for CI/CD

In this section, we will mention some of the tools I have seen during my career, with no preference for any of them specifically. We are sure (in fact, we hope) that you are able to find alternatives and consider the best ones for you. These are just examples, and we are sure there are many other interesting ones.

We have divided the various tools into sections, but nowadays most tools allow for several uses.

## Programming languages

Most of the CI/CD tools would use some programming language to specify the instruction of what the deployments need to do at every step, such as running tests, installing commands, and so on. Each favors one programming language over another, but in my mind, there are some that you should be familiar with. In the following table, we provide a summary of interesting languages to get familiar with. The new concepts or words in the table will be reviewed in the following sections.

| Name | Description | Why is it important? |
| --- | --- | --- |
| YAML ain't a Markup language (YAML) (see [1] in the *Further reading* section) | YAML is a data-serialization language. | It is largely used for configuration files and is one of the first languages used for deployment step definition. It is also used in docker-compose files. |
| XML/JSON [2] | Extensible Markup Language (XML)  JavaScript Object Notation (JSON) | Commonly used for requirements definition. They are an alternative to YAML but less human-readable. |

| Bash/Batch [3] | Bash is a Unix shell and command language. Batch/shell scripts are the equivalent for Windows. | It is frequent to have one or more operating system (OS) scripts called from the deployment steps to perform some operation due to its complication or its OS-related specifics. |
|---|---|---|
| Typescript [4] | Typescript is built upon JavaScript and provides strong typed definitions for methods and objects. | It is used for a variety of purposes, for example, it is also used to write VS Code extensions. |
| Go [5] | Go is an open source, statically typed, compiled programming language that is similar to C, with features that are found in newer languages. | It is a higher-level language that allows for lower-level operations, popular for scripting files related to CI/CD. |

Table 1.1: Programming languages related to CI/CD

## CI/CD tools

These tools have the main capabilities required to automate CI/CD pipelines. The tools provide security, user creation, SSH key generation, installing plugins, and maintenance of log files and backups. Let us have a look at some of these tools in more detail:

- **Gitlab** *[6]*: This is a web-based DevOps life cycle tool providing a Git repository manager. It also provides features such as issue-tracking, wiki, and a CI/CD pipeline. It uses YAML for configuration and step definition.

- **Jenkins** *[7]*: This is an open source automation server. Jenkins supports building, deploying, and automating projects. It comes with a straightforward UI to perform configuration and can be integrated with other systems, for example, Gitlab. It also provides a nice interface for test automation and issue tracking.

  My recommendation is to use it for CI/CD only if you are planning on using it for something else.

- **Ansible** *[8]*: This is an open source IT engine used to automate application deployments, service orchestration, cloud services, and other IT tools. As before, it can be used for CI/CD if you are planning to give it other uses.

- **TeamCity** *[9]*: This is a build management and CI server from JetBrains. As before, it also allows for other cool features, such as test automation, and it has plugins for version control, and issue tracking.

- **Azure Pipelines** *[10]*: Azure Pipelines is Microsoft's cloud-hosted code pipeline.
- **AWS CodePipeline** *[11]*: AWS CodePipeline is Amazon's web service code pipeline tool.

## Packaging tools

These tools allow you to deliver software by packaging what is needed for the deployment (software, libraries, and configuration files) in isolated containers allowing for faster reproduction and recreation. Installing and configuring the environments would be too manual if it weren't for these. They also help us with the scaling and management of the applications:

- **Docker** *[12]*: This is a **platform-as-a-service** (**PaaS**) product for packaging software in containers
- **Kubernetes** *[13]*: This is an open source system to orchestrate containers
- **Amazon Elastic Container Service** (**Amazon ECS**) *[14]*: This is a fully-managed service to orchestrate containers

## Performance monitoring tools

We have highlighted several times across the book the importance of the performance tracking and debugging capabilities of the applications, especially when they are distributed across different servers or in the cloud. Some of the tools mentioned previously already come with performance and monitoring capabilities, but I think it is worth mentioning Elastic and the ELK stack:

- **Elasticsearch** *[15]*: This is a distributed search and analytics engine for all types of data
- **Logstash** *[16]*: This is an ingestion pipeline
- **Kibana** *[17]*: This is a visualization tool

This stack allows you to aggregate and analyze logs and create a visualization for easier and faster tracking of issues, although they are not the only ones around that can help you with this, they are very easy to learn and to get you started with performance monitoring.

See *[18]* for AWS solutions such as **AWS CloudWatch**, **CloudWatch ServiceLens**, and **AWS X-Ray** and *[19]* for Azure's **Application Insights**.

## Version control systems

**Version control system** (**VCS**) tools allow you to save different versions of your code, make branches, and make it possible for different developers to work on the same piece of code. Among them, we can find tools such as **Git**, **Subversion** (**SVN**), **ClearCase**, and **Concurrent Versions System** (**CVS**). **Visual Studio Team Services** (**VSTS**), previously known as **Team Foundation Server** (**TFS**), is also worth a mention, especially when working with Microsoft technologies.

## Workflow applications

These applications allow you to mix and match different steps and different technologies and handle them at a higher level. You could combine repositories, containers, CI/CD pipelines, and functionality and performance tools.

The following are some examples:

- **GitHub Actions** *[20]*: This provides a way to implement complex CI/CD functionality directly in GitHub by initiating a workflow on any GitHub event.
- **Terraform** *[21]*: This is an open source **infrastructure-as-code** (**IaC**) software tool. You can define data center infrastructure using configuration files written in JSON or **HashiCorp Configuration Language** (**HCL**) (a language specific to Terraform).

All these tools are wonderful and really help us with our CI/CD needs. However, sometimes, the best way of understanding how something work is by re-creating it from zero or doing at least something equivalent. In the next section, we are going to see how.

# A basic CT example

In this section, we are going to investigate how it is possible to automate CT from zero, without any tools (other than Git for source control). For simplification, we will not cover CI, but we challenge you to try it yourself as a learning project. Feel free to review the definitions of CI/CD at the beginning of this chapter again if you struggle to understand what would be different.

In our examples, we will be using vstest to create and execute tests, as this can be easily called from batch/cmd and from other programs. This tool generates output that we can add to a text file for later analysis. Then, we will showcase how to create a YAML file to run the entire process to get you familiar with YAML and how it links all the tools together. Finally, we will use Python to call and execute the YAML file. This way, you can see how different tools with different programming languages can interact.

I hope that the concepts explained in this section help you improve your CI/CD systems or automate smaller tools for your team.

To perform a basic CD, we will create an automation that can run some unit tests using Visual Studio. We will provide these tests in a Git repository. Then, we will define a YAML file with all the instructions needed so we can add and change as we need, and a Python file to parse the YAML file and execute such instructions. Finally, we will check how to get notifications for the failed test cases.

## 1. Running tests programmatically

Running tests programmatically is part of the concept of *automating the automation* that we have already covered in *Chapter 1, Introduction – Finding Your QA Level* and *Chapter 5, Testing Automation Patterns*. Instead of manually launching them, we can now set up a system to do so, in this case, the system will be our own CD.

Given we have our tests already defined in Visual Studio (see `https://github.com/PacktPublishing/How-to-Test-a-Time-Machine/blob/main/Chapter06/UnitTestProject1.dll`), we need to execute them on our machine.

In the Visual Studio installation, there is a tool to execute test cases called `vstest.console.exe`. In Windows, it is generally located under `C:\Program Files (x86)\Microsoft Visual Studio version\Common7\IDE\CommonExtensions\Microsoft\TestWindow`. Please note that the version will be different depending on your version Visual Studio, it might contain some sub-folders, and this tool might be outdated by the time you read this book, so feel free to change this bit to whichever tool is needed for you to run your unit tests programmatically.

Then we need to point to our `dll` test code. This is generally stored under your test project's `bin\Debug` folder. We have provided one in our Github repository, so you could use that one. Keep in mind all the tests will fail on this one.

Finally, we can give parameters to `vstest` to run certain tests, grouping them. By default, all the tests will be executed, so we are not passing any further parameters. All in all, the way to execute these tests from a command line or terminal would look like this:

```
"C:\Program Files (x86)\Microsoft Visual Studio YourVSVersion\
Common7\IDE\CommonExtensions\Microsoft\TestWindow\
vstest.console.exe" How-to-Test-a-Time-Machine\Chapter06\
UnitTestProject1.dll
```

The next step is to save the test results.

## 2. Logging

While it is very nice to be able to execute tests programmatically, it is of no use unless we can record the results. Again, from a command line or terminal, it is just a matter of sending the output of the call into a text file, as shown in the following lines:

```
"C:\Program Files (x86)\Microsoft Visual Studio YourVSVersion\
Common7\IDE\CommonExtensions\Microsoft\TestWindow\
vstest.console.exe" How-to-Test-a-Time-Machine\Chapter06\
UnitTestProject1.dll > testOutput.txt'
```

Now we have a way of executing the Visual Studio unit tests programmatically (for a Windows system). We can use this call as an instruction from the YAML file.

## 3. Creating a YAML file

Although this step is not needed for a successful CD automation, since YAML files are so common in CI/CD systems, we thought it would be useful to see how they work as a configuration file for these systems. Note that for different technologies, the YAML file has to match the schema. In our case, we can make it a little more freely, as in the following example.

The main benefit of using a file to set up the instructions, rather than having them hardcoded, is that adding new instructions becomes much easier and seamless.

In the following example, we create a YAML file with instructions to clone some code and execute the tests against that code. We are keeping it simple, but we will likely want to clone and/or pull the test code as well. If the pull has a merging issue, this script will fail:

**Instructions.yml**

```
Steps:
- task:
  name: cloneRepo
  log: Cloning repository
  instruction: git clone https://github.com/
PacktPublishing/How-to-Test-a-Time-Machine.git && git init &&
git fetch
- task:
  name: pullingRepo
  log: Pulling repository...
  instruction: cd How-to-Test-a-Time-Machine && git pull
- task:
  name: executeTests
  log: Executing tests...
  instruction: '"C:\Program Files (x86)\Microsoft Visual
Studio YourVSVersion\Common7\IDE\CommonExtensions\Microsoft\
TestWindow\vstest.console.exe" How-to-Test-a-Time-Machine\
Chapter06\UnitTestProject1.dll > How-to-Test-a-Time-Machine\
Chapter06\TestOutput.txt'
```

Next, we will parse this file so we can execute the commands. Please make sure to put your vs version in YourVSVersion if you are using this code.

## 4. Executing the commands

In the following Python example, we will read the YAML file and iterate through its steps, logging what we are doing as well. This way, we don't need to hardcode instructions, and we can add more later by simply modifying the YAML file rather than creating extensive code:

**test_execution.py**

```python
from yaml import load, safe_load, Loader
import os
from git import Repo

with open('instructions.yml') as stream:
    data = safe_load(stream)
    for step in data['Steps']:
        print(step['name'])
        print(step['log'])
        os.system(step['instruction'])
```

Now we have an automatic system that runs commands on demand. If the tests get modified, we will pull the latest code and execute them automatically with this code. Of course, we should also pull the latest feature code as well. We can have different YAML files for different systems, so we can execute the one for our OS, rather than having more complicated Python code.

If we wanted to add CI, we have a similar way of automatically adding the new code to the GitHub repository every time there is a change on it.

As mentioned before, while this is good to have, it won't be much use unless we also get notified about what has happened. In the next section, we will see how to use the logs we have recorded to get different notifications.

## 5. Getting notified

There are several ways we can get notifications if anything goes wrong. For CI/CD, it is important to block the delivery process if that has happened. Lucky for us, the CI/CD tools that we have reviewed in this chapter have that functionality already built in. However, most of the time, you must refresh the output page to check whether the build has passed or failed.

Sometimes, we might want to be alerted in a more direct way; for example, by receiving an email with a summary of the test execution or a comparison with previous executions, so we have a clear picture of everything that is going on. If we want to save space in our inbox and we are using a messaging system such as Slack, we might want to receive a quick Slack ping when the build has finished instead.

If something goes wrong, we might prefer to receive an SMS, a call, or a WhatsApp message for a quicker follow-up action.

The first step is to analyze the logs that we have received. We could use any programming language to do this. For example, the following Python script will analyze the log file (testOutput.txt) using regular expressions and search for failed strings:

## log_analyzer.py

```
import re
print("starting")
file_object  = open("testOutput.txt", "r")
x= re.findall("Failed: .", file_object.read())
if(len(x) >0):
    failedTests = x[0][8:]
    print(failedTests)
```

We can call this piece of code from the YAML file (python How-to-Test-a-Time-Machine\ Chapter06test_execution.py). If we execute this code after the tests run, we will now have a way of understanding the failures and passes. We can also use Python's subprocess library (with the subprocess.run call) to execute some code here to perform the notification actions mentioned previously.

We will learn more about getting notifications by email, text, WhatsApp messages, calls, and Slack in *Chapter 11, How to Test a Time Machine (and Other Hard-to-Test Applications)*.

## Summary

In this chapter, we covered CT and other different DevOps concepts that we should be familiar with to ensure quality across the entire development and deployment, and thus proving how CD really is all about automation. We will learn a bit more about how everything interconnects, including some diagrams, in *Chapter 11, How to Test a Time Machine (and Other Hard-to-Test Applications)*, in the *Architecting a system* section.

In the next chapter, we will look at something challenging but nonetheless needed to understand the rest of the chapters fully, the mere mention of which might initially intimidate some of you when reading it, but trust me, we will make it as fun as possible. It is also recommended that you read it before the next chapters as it will lay out important concepts that you may need for them. If you need to, take a deep breath, and then keep on reading…

# Further reading

- [1] Official website: `https://yaml.org/`

- [2] For XML, check out: `https://www.w3.org/standards/xml/core, for json check https://www.json.org/json-en.html`

- [3] A good compilation of links to learn Bash: `https://www.redhat.com/sysadmin/learn-bash-scripting`. To learn something more advanced than batch for Windows scripting, you can find a compilation of PowerShell commands here: `https://devblogs.microsoft.com/scripting/table-of-basic-powershell-commands/`

- [4] Official website: `https://www.typescriptlang.org/`

- [5] Official website: `go.dev`

- [6] Official website: `https://gitlab.com/gitlab-com`

- [7] Official website: `https://www.jenkins.io`

- [8] Official website: `https://www.ansible.com`

- [9] Official website: `https://www.jetbrains.com/teamcity`

- [10] More here: `https://learn.microsoft.com/en-us/azure/devops/pipelines/get-started/what-is-azure-pipelines?view=azure-devops`

- [11] Official website: `https://aws.amazon.com/codepipeline/`

- [12] Official website: `https://www.docker.com`

- [13] Official website: `https://kubernetes.io`

- [14] Official guide: `https://docs.aws.amazon.com/AmazonECS/latest/developerguide/Welcome.html`

- [15] More here: `https://www.elastic.co/elasticsearch`

- [16] Official website: `https://www.elastic.co/logstash`

- [17] Explore more here: `https://www.elastic.co/kibana`

- [18] See `https://aws.amazon.com/what-is/application-performance-monitoring/` for a deeper explanation of application performance monitoring and AWS solutions for them

- [19] See `https://learn.microsoft.com/en-us/azure/azure-monitor/app/azure-web-apps` to learn how to do application performance monitoring with azure

- [20] Official guide: `https://docs.github.com/en/actions`

- [21] Official website: `https://developer.hashicorp.com/terraform`

You can learn some more about scheduling automation and CI/CD on Packt's website, for example:

https://www.packtpub.com/product/bmc-control-m-7-a-journey-from-traditional-batch-scheduling-to-workload-automation/9781849682565?_ga=2.88165212.658457769.1671995044-2068911769.1671995044

https://www.packtpub.com/product/devops-project-2022-cicd-with-jenkins-ansible-kubernetes-video/9781803248196

# 7
# Mathematics and Algorithms in Testing

In the previous chapters, we learned about different types of testing across the test pyramid. In *Chapter 5*, *Testing Automation Patterns*, we deepened our knowledge of UI/E2E testing, which, for better or for worse, tends to end up being the biggest focus for many applications. In the previous chapter, we talked about continuous testing and other DevOps concepts that we should know about to make sure that the quality stands across the development and deployment aspects.

Before we look at more challenging topics, we need to take a little break and look at some mathematical foundations and concepts. In this chapter, we will look at the mathematics that can help you with your tests.

Mathematics is sometimes overlooked in computer science, especially in the testing area. However, having knowledge of and, arguably, love for mathematics would put you forward in your career and improve the quality of the tools that you create Every computer science degree has an important number of math-related courses, and that is for a reason.

A lot of people hate them and I used to be one of them. I could not understand why I needed them so much in university since I had been writing code since I was very young without using them explicitly. But I began to learn to love them, and the more I loved them, the more I understood how helpful they are. In this chapter, I hope to inspire some of that love for them in you too.

In this chapter, we are going to cover the following main topics:

- Algorithms, mathematics, and testing
- Understanding the role of data science in testing
- Reviewing some techniques for test case analysis
- The early approach to AI – it's all thanks to mathematics

# Technical requirements

Some degree of programming skill is recommended to get the most out of the examples provided. In this chapter, we will use a variety of programming languages, mainly C#, Python, and JavaScript.

We recommend reviewing and working with different languages as a self-growth exercise. We have provided an implementation for other languages in this book's GitHub repository: `https://github.com/PacktPublishing/How-to-Test-a-Time-Machine/tree/main/Chapter07`.

While this chapter was written with the QA/SDET role in mind, as applications shift left, developers may find this one interesting. Furthermore, if you are trying to get developers more involved in testing, this is the chapter you will want to show to them as it will hopefully trigger their "building" instincts and curiosity.

# Algorithms, mathematics, and testing

In this section, we will learn how and why mathematical algorithms can help with the quality of applications and their testing.

Mathematics is all around us and as it happens with many things, once you start gaining interest in it, you will see it everywhere: in the shape of plants and fruits whose leaves grow following mathematical series to find the highest amount of sunlight, in music and harmony, in the stars, in atoms, and more.

As we mentioned in the introduction, it is perfectly possible to solve programming challenges without the explicit use of mathematics. Note the word *explicit* here; we implicitly use mathematics in algorithms to calculate the size of data structures, perform complex calculations, etc.

However, understanding the mathematical approach could help you improve algorithms and code complexity and reduce the use of resources and data structures. This could give you an advantage while reviewing someone else's code, when writing it yourself (whether it's feature code, tool code, or test code), or even when writing testing for those algorithms.

Knowing which solution is the best is highly related to your system and the best procedure for finding one is to plan for a few solutions as there are times that some might work better up to a certain value. It is important to be able to calculate and compare time and space complexities and also take development time into account; sometimes, a quick solution is preferred over a perfect one. Creating benchmarks to measure solutions is not part of this chapter's goal, but we recommend you research and practice this on your own.

You could reach a solution for an algorithm in two ways: by reasoning or by knowledge. Having a previous knowledge of the solution is faster than having to reason it from 0, but nobody is expected to know everything, and being able to reason is an important skill that you could develop through practice.

I would love to give you some guidance about what algorithms you should learn or have handy, and what sort of mathematical concepts would help you throughout your career, but that would take several books and it would be too specific to each person. Therefore, here, I have provided a list of subjects that could be useful for you and specified when each of them would be more useful. Please feel free to learn more about these and add your own ones if you find they have been or could be helpful:

- **Polygons**: This concept is especially interesting if you deal with graphics, graphical representations, and polygons. Also, if you deal with many mathematical functions, this knowledge might be helpful for you to visualize the following functions:

    - Areas of the polygons

    - Volumes of geometric bodies

    - Segments of spheres

    - Angles of circumferences

    - Segments (Thales' theorem)

- **Graph representations**: Similar to polygons; if you are dealing with 3D graphics, you can visualize mathematical functions. You might come across them in artificial intelligence:

    - 3D representations; for example, Euler vectors and Quaternions

    - Graphs of functions (constant, decreasing, increasing, second grade, convexity, and so on)

    - Max and min functions

    - Sine and cosine theorems

- **Signals**: These are important when you're dealing with hardware and creating hardware simulators; for example, a Fourier transform.

- **Logic and algebra**: These are especially important concerning data science and AI, but might be useful in other cases:

    - Matrix handling; for example, dot product, multiplication, Hadamard product, determinant, invertibility, linear dependence, etc.

    - Probability

    - Variance

    - Standard deviation

    - Entropy

    - Gaussian density

    - Newton's method

- **Algorithms**: Algorithms are mostly for everything to do with computing, especially when dealing with trees, graphs (such as JSON objects or DOM manipulation), and complex data structures:

  - Dijkstra

  - Backtracking

  - Dynamic programming

  - Sorting algorithms (quick sort, merge sort, and so on)

  - Graph and tree traversal

  - Bit manipulation

- **Other concepts**: Here are some other concepts that might be useful across several problems:

  - Digital root

  - Error-correcting code (especially for cryptography)

  - Data manipulation (arrays, strings, linked lists, stacks, queues, and more)

These topics are not always easy to remember and it is good practice to refresh yourself on them every so often. Sometimes, you will be dealing with a problem and you will realize it could easily be solved by using a data structure or algorithm. Sometimes, you will find a library that does that for you, but unless you understand that it could be solvable with it, you will never think of using that library in the first place.

While you could find any of these concepts during your career, when it comes to testing, I would say one of the most important ones is graph traversals, especially if you are considering creating some sort of application crawler for automatic discovery testing.

> **Application crawler**
>
> An **application crawler** is an automatic system that iterates throughout the different parts of an application, generally randomly, to find issues or explore the application.

## Graph traversals

One of the algorithm concepts that we might be interested in, especially in the testing area, is the different ways of traversing a graph (or a diagram showing a relationship between concepts). In this section, we will review how and why we are interested in this and how to perform such traversal specifically for an application crawler, to perform exploratory testing of such applications.

A graph is composed of two main sections: the nodes (the concepts) and the arrows, arcs, or connections between them.

Most applications' user interfaces can be thought of as some sort of graph, where the pages/views/screens/levels are the nodes and the elements inside them that navigate to another view are the connections between the nodes. This vision of an application allows us to use concepts such as graph traversals and benefit from already proven and stable algorithms:

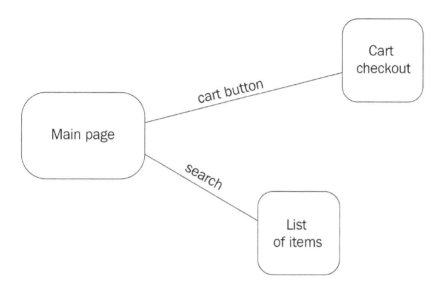

Figure 7.1: Example of a website shown as a graph

We are going to refer to the screen/pages/levels as views from now on. *Figure 7.1* would, therefore, have three *views* which are represented by three nodes of the graph. For the same figure, we see two represented arcs, which would be the two buttons that would take us from one view to the next.

There are two (main) ways of traversing graphs:

- We can visit a view that connects to our current view, go back, then visit another view connected to our current view until there are no more unvisited views connected to our current view. Then, we would go to the first one that we visited and repeat this process for other connections to the visited views. Before going to another level, we would first explore all the views that were linked to our first screen. This is called breadth-first. In *Figure 7.2*, we can see that, first, we visit our starting page, called **Main page**; then, we go to our next page, **List of items**, then travel back to another of the connections to **Main page**, such as **Cart checkout**, until we exhaust those connections before going to the next level:

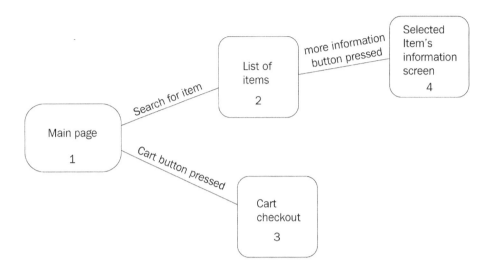

Figure 7.2: Breadth-first example – the numbers represent the order in which the visit takes place

- We can visit a view that links to our current view and keep going forward to another view available from the previous one until there are no more links left to visit in that depth. Then, we go back until we find another path that is yet to be explored and explore it in the same way. This is called depth-first. In *Figure 7.3*, we can see that, first, we navigate to the main page; then, we navigate to the second page, **List of items**, and keep going deeper into that route to **Selected item's information screen** before going back. Since **List of items** does not have any more connections, we go back again. **Main page** does have a connection, so we visit **Cart checkout**:

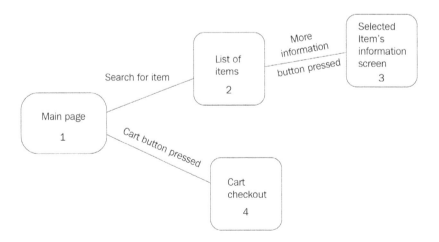

Figure 7.3: Depth-first example – the numbers represent the order in which the visit takes place

The breath-first method could be useful if we want to deeply test a particular view of an application. If our application does not have a lot of different actions per view but has a long list of succeeding views, depth-first would be more convenient. However, as we mentioned previously, the idea of a general application crawler is for it to be *random*, and both systems (breath-first and depth-first) might end up covering the same areas constantly.

Both systems are very exhaustive, and while they could be of use to explore an application, they might not be very close to user behavior. If your typical user behaves in a depth-first or breadth-first way, or the way your application is constructed would benefit from using either system, feel free to use that iteration, especially if your application can be fully covered in a reasonable amount of time (for which you would not need much randomizing). For eample, if your application is a game with levels that do not repeat, using depth-first approach would mimic better the user experience. Both algorithms are widely known and you should be able to find plenty of documented resources about them.

Alternatively, if we have a fully random system, we can miss important parts of the application while repeating the same tests over and over. You should consider adding weight to the different parts; we will see how in *Chapter 8, Artificial Intelligence is the New Intelligence*.

Since we want you to get the most out of this book and cover the parts that are not generally covered elsewhere, we are going to create a new randomized traversal that uses the same idea of the two systems that we discussed previously: keeping track of visited and to be visited nodes.

There is another addition to our code, which is to interact with everything interactable. Sometimes, these interactions will not drive us to another view. For simplification, we are showcasing the code for a web crawler, but it could be adapted to another sort of application.

Every time we visit a view, we must collect all actionable items for that view and the view URLs or identifiers. Those URLs will be saved into a visited map, alongside the number of interactions still available on it. We must randomly pick an interaction from our current view; if none are available, we must go back or switch to another view if going back is not an option.

We should also keep track of the time spent on the algorithm. The code will finish if everything is visited, if we run out of time, or if we found any issues (for example, an error code such as 404 or some sort of resource leak).

Finally, in websites, we might have links to an external source that is not part of our website. We should make sure we can navigate there but that this is not saved as another view. This can be done by making sure the saved views are all under a specific URL domain.

We can summarize all of those actions in the following pseudocode, in which we will explore the main parts of our algorithm:

## Pseudocode algorithm for exploratory random testing:

```
Insert URL homepage into possible list and create a node.
Set to visited.
```

```
Add all possible URLs to go from it / to act upon.
Select a random href element with that URL and set to acted.
Click on the element to verify it is clickable and no 404
happens.
Repeat for each URL redirection.
If a new tab opens, close it.
```

Let us implement an example exploratory test for `https://www.packtpub.com/website` (remember to use `pip install` for any missing libraries on execution):

1.  First, here are the imports that we will need. Note that `node_class` is a class that we will be creating later; it is not to be confused with any library by that same name:

### simple_exploratory_random_testing.py – imports

```python
from selenium import webdriver
from selenium.webdriver.common.by import By
import requests
import sys
import random
import node_class
from selenium.webdriver.chrome.service import Service
from webdriver_manager.chrome import ChromeDriverManager
from selenium.webdriver.support.ui import WebDriverWait
from selenium.webdriver.support import expected_
conditions
from selenium.webdriver import ActionChains
```

2.  This is the main class, which will perform the exploration. We have a constructor in which we initialize the driver, the number of actions we will perform, and the beginning URL for the graph (in our case, `packtup.com/free-learning`). Then, we start the exploratory testing and finalize the driver afterward:

### simple_exploratory_random_testing.py – main class

```python
class SimpleExploratoryRandomTesting:
    def __init__(self):
        driver = webdriver.Chrome(service=Service(
            ChromeDriverManager().install()))
        driver = webdriver.Chrome(
```

```
            ChromeDriverManager().install())
        self.top_level = 10
        url = "https://www.packtpub.com/free-learning"
        driver.get(url)
        handle = driver.current_window_handle
        node = node_class.NodeClass(url)
        node.add_handle(handle)
        self.simple_exploratory_random_testing(
            node, [], 0, driver)
        driver.close()
```

3.  In the following method, we are performing the actual graph traversal. Here, we keep track of the current level and compare it to the top one. We must add the current node to the visited list, check for potential errors on the URL (with API check), get all the possible actions (or `hrefs`) on the current view, and iteratively pick random ones from the not visited views and explore them:

### simple_exploratory_random_testing.py – main class – exploratory testing method

```
def simple_exploratory_random_testing(
self,node, visited, current_level, driver):
    """ method to start the exploration """
    current_level = current_level + 1
    opened_tabs = len(driver.window_handles)
    if current_level >= self.top_level:
        sys.exit("Max visits reached")
    visited.append(node)
    status_code =
        requests.get(node.url).status_code
    if status_code < 200 or (
        status_code >= 400 and status_code < 980):
        # hard exit to signify an issue on the url
        sys.exit("Error on the url " + node.url)
    # get all actions first time in the node
    try:
        driver.switch_to(node.window_handle)
        driver.find_element(
```

```
              By.XPATH,
              '//a[@class="accept_all"]').click()
except Exception:
    print("no cookies found")
if node.count == -1:
    node.count=0
for a_tag in driver.find_elements(
    By.TAG_NAME,'a'):
    # we only get the ones with packtput.com
    # on them to be new nodes
    if "packtpub.com" in node.url:
        node.count=node.count+1
        href = a_tag.get_attribute('href')
        if href and not href.startswith("#"):
            node.actions[href]=
                a_tag.get_dom_attribute('href')
# iterate through the actions randomly and
# visit them
while node.count > 0:
    # randomly get one - then append to acted
    # and remove a count
    suburl = self.random_get_one(node.actions,
                               node.acted)
    node.acted.append(suburl)
    node.count = node.count - 1
    # create a new subnode
    subnode = node_class.NodeClass(suburl)
    if subnode not in visited:
        self.try_click(
            '//a[@href=
            "'+node.actions[suburl]+'"]',
            driver, driver)
        # repeat the process for this subnode

        self.simple_exploratory_random_testing
```

```
                    (subnode, visited, current_level,
                        driver)
                driver.implicitly_wait(5)
                subnode.add_handle(
                    driver.current_window_handle)
            # verify we havent reached the max of
            # actions taken, so the algorithm does not
            # take too long
            if current_level >= self.top_level:
                sys.exit("Max visits reached")
            # close any opened tabs
            if len(driver.window_handles) > \
                    opened_tabs:
                driver.close()
                driver.switch_to().window(
                    node.window_handle)
        else:
            driver.get(node.url)
            driver.implicitly_wait(5)
```

4.  In the following piece of code, we will get a random URL from those available in the actions map that have not been acted on already:

### simple_exploratory_random_testing.py – method to get a random suburl

```
def random_get_one(self, actions, acted):
keys = list(actions.keys())
    count = len(keys)
    ransel = random.randrange(count)
    suburl = keys[ransel]
    while suburl in acted:
        ransel = random.randrange(count)
        suburl = keys[ransel]
    return suburl
```

5.  In the following piece of code, we are trying to click on the `href` element to move to the next view:

### simple_exploratory_random_testing.py – method to try to click in the href elements

```python
def try_click(self, xpath, driver, parent):
    try:
        element =
            parent.find_element(By.XPATH,xpath)
        ActionChains(parent).move_to_element(
            element).perform()
        driver.implicitly_wait(5)
    except Exception:
        print("Could not find the xpath" +
        ", it is possible that this object" +
        " is not available anymore")
```

6.  The following code shows the call to the main entry point of the exploratory testing class:

### simple_exploratory_random_testing.py – main entry point

```python
SimpleExploratoryRandomTesting()
```

7.  The following code is used to define the graphs by defining the node and the arcs/connections (also known as actions). We also keep what has been acted on before and the number of available actions left:

### node_class.py

```python
class NodeClass:
    url=""
    Window_handle=""
    actions={}
    acted=[]
    count=-1
    def __init__(self, url):
        self.url = url
            def add_handle(self, window_handle):
                self.window_handle = window_handle
```

There are some problems and considerations we should keep in mind regarding our exploratory testing solution:

- We always close cookies since the pop-up was causing issues for us.

- `href` elements could be hidden (we use a JavaScript code to click the elements, but maybe we should click the dropdowns or, depending on the type of parent element, perform any other actions).

- We missed some actions, such as sliding and drag and drop.

- We might miss visual issues, for which it would be great to accompany them with a visual testing tool. Additionally, if you have a list of expected objects for the view, you could also check they all are visible.

- This code does not work for things such as form completion with mandatory fields. For those cases, consider a random test crawler.

- Note that some URLs might come with strange status codes; make sure your class fails on the right ones.

- The example URL has some randomized `href` elements, so we skip it if we do not find the elements to simplify the code. However, the ideal solution will make sure links are clickable. A random solution might not be the best for this particular case.

> **Random test crawler**
>
> A **random test crawler** is an automatic system that executes the different tests of an application randomly.

- In the code, we are effectively only measuring 404s and skipping unclickable links, so it might be as good to have an API crawler instead, which makes sure the calls to the APIs are correct. This would be easier to do than the UI version showcased.

- Sometimes, some objects behave differently if clicked more than once. If that is our case, we might want to have a counter with each interaction to make sure we reuse them or think of another solution, such as clicking every object twice if it is still present.

- There are some behaviors that this code would not cover, such as adding things to a cart before purchasing. Think of strategies to add this to your crawler.

- For non-web applications, getting a differentiable URL is not trivial, but doable, as we will explore in *Chapter 11*, *How to Test a Time Machine (and Other Hard-to-Test Applications)*.

- For dynamic websites, the interactable objects might change. If that is the case, you should treat each group of objects as a different view, similar to what we will do for the non-trivial views mentioned in the previous point.

- To measure resource leaks, we should include and review metrics, which we will review in the next section.

- In the examples shown in *Figure 7.2* and *Figure 7.3*, we can see how we travel to a different view when we click more information on an element. This URL will likely change per item; however, we may prefer only traveling to one of the items rather than spending a long time visiting them all, as that would give us better information.

- Finally, if there are some views of the application that we want to cover more profoundly, we could assign weights to the random function so that we retrieve more items from these parts.

All of these items are considerations and challenges that your crawler might need to overcome.

Problems such as the one mentioned at the beginning of this section might seem difficult to solve, or you might think you would never find them in the test area, but they are quite frequent everywhere and they get easier and easier the more you practice them.

In the next section, we will cover some data science concepts and understand why we need them in testing. One of the potential uses for them could be to try to analyze the data coming from the web crawler or some other source and figure out potential common issues in our application.

## Understanding the role of data science in testing

The data science field uses scientific mathematical methods to understand relationships and extract knowledge from data so that we can use this knowledge to our benefit. This, of course, applies to testing. Therefore, mathematics gives us methods that we can use to find relationships between the different data of our system and value performance and other metrics.

In the previous chapter, we discussed how important it is to use metrics and log the right information in our system. Most of this information comes in one way or another from testing. However, even though it is important to extract this information, being able to make sense of it is just as important. This way, we can improve the system and provide better-quality applications to the users.

Your company may have a specialized data scientist that could help you achieve this purpose, but I believe that this is a field you should strive to understand and be comfortable with as well. The reason for this belief is that, in the majority of cases, the data scientist would need an expert to make sure their calculations and algorithms are working as expected.

Many different bits of data could be helpful for your application – for example, the parts of the application that cause more bugs. It is important to measure the different components, client feedback, and even data related to the development and testing of the application (for example, how many test cases are automated by each team) as much as possible.

Another example that could be very important for us is the one mentioned in the previous section: finding areas of the application that would need more testing. While we will see how to achieve this in the next chapter, it would be not possible to do this unless we have data that can help us understand what areas of the application fail more frequently.

There are many applications nowadays that allow us to visualize such data and generate automatic dashboards or provide our own. However, I find it important to be able to understand and create personalized dashboards as that might come in useful at some point.

## Creating a dashboard

Creating a dashboard could be useful to showcase different sides of your application, as we mentioned previously. Furthermore, we can have a visualization of the nodes and arrows of our application as it has been traveled by the crawler that we created in the previous section.

If you are creating a dashboard, chances are that you would like to share it with other people, at least in your team. JavaScript could be the language for this. A good resource for creating dashboards is JavaScript's **3D.js** *[1]* library. You can also consider using some other language, such as Python's **matplotlib** *[2]* library, to generate an image on a server and serve it as a response of a URL, but that would be more advanced, so we will review D3.js for now, which is much easier to use.

Now, you can add the reference script shown in the code examples provided. D3.js provides many different functioning charts and other visualizations that you can use directly. To install it, make sure you have installed and initialized NPM (npm init) and D3.js (npm install --save d3) in your project.

For example, using the aforementioned library, you could easily create a pie chart that represents the state of the automation for a team – for example, the percentage of test cases that are automated, the percentage of test cases that are not automated but we want to automate, and the percentage of test cases that are not automated and we are not going to automate. Alternatively, you could show the percentage of test cases passing or failing. We could have a graph per team or/and per delivery, so we could compare two views. Let us get started:

1.   First, we will create a simple HTML document to hold the chart:

### PieChartExample.html

```
<!DOCTYPE html>
<meta charset="UTF-8"/>
<html>
<head>
<script type = "text/javascript" src = "https://d3js.org/
d3.v4.js"></script>
</head>
```

```
<body>
<div><br></div>
<chart></chart>
<script src="pieChartExample.js">
</script>
</body>
</html>
```

Note that in this document, we are linking a script called `pieChartExample`, as shown in the following code, which contains the logic for creating the graph.

2.  Now, we will define the sizes that we want our figure to have and append the drawing to the div created in the preceding HTML (called `<chart>`):

### pieChartExample.js – defining the sizes and appending the drawing

```
var width = 400
    height = 400
var outerRadius = Math.min(width, height) / 2 - 30
var svg = d3.select("chart")
  .append("svg")
    .attr("width", width)
    .attr("height", height)
  .append("g")
    .attr("transform", "translate(" + width / 2 + ","
                                    + height / 2 + ")");
d3.json("testcases.json", function(d) {
    createChart(d);
});
```

3.  Then, we must create the pie chart and add the colors:

### pieChartExample.js – logic to create the chart

```
function createChart(rows) {
var data = rows;
var color = d3.scaleOrdinal()
  .domain(data)
  .range(["#A8DBA8", "#CFF09E", "#79BD9A"])
  // add more colours if we have more cols
```

```
var pie = d3.pie()
  .value(function(d) {return d.value; })
var pie_data = pie(d3.entries(data))
```

4.   Now, we must add the data and the arcs to this pie chart:

## pieChartExample.js – logic to create the data

```
// create data
svg
  .selectAll('div')
  .data(pie_data)
  .enter()
  .append('path')
  .attr('d', d3.arc()
    .innerRadius(0)
    .outerRadius(outerRadius)
  )
  .attr('fill', function(d){ return(color(d.data.key)) })
  .attr("stroke", "black")
  .style("stroke-width", "2px")
  .style("opacity", 0.6)
```

5.   Finally, we must add the labels to the pie chart:

## pieChartExample.js – logic to add labels

```
//add labels
svg
  .selectAll('div')
  .data(pie_data)
  .enter()
  .append('text')
  .text(function(d){ return d.data.key})
  .attr("transform", function(d) {
      return "translate(" + d3.arc()
                              .innerRadius(0)
                              .outerRadius(outerRadius).
```

```
centroid(d)
                                    + ")";   })
   .style("text-anchor", "middle")
   .style("font-size", 20)
}
```

The testcases.json file could be automatically generated from our test data (for example, from the TestRail API, as we saw in *Chapter 3, The Secret Passages of the Test Pyramid – The Middle of the Pyramid*). It would look like this:

## Testcases.json

```
{"automated":60,"not-automated":30,"wont-automate":10}
```

The chart will look like this:

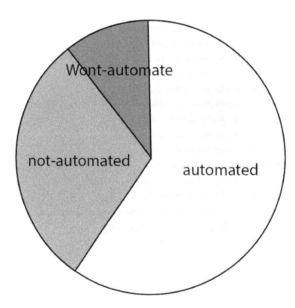

Figure 7.4: Pie chart result

This is a basic example, but with some JavaScript understanding, you could create animations between teams or projects, highlighting slices of the chart, and so on.

For example, you could use a pyramidal chart from this library *[3]* to easily represent the test pyramid of the team. This can be seen in the following example. For simplification, we are adding the script within the HTML, but we recommend that you have two files instead:

## pyramidalExample.html

```html
<!DOCTYPE HTML>
<html>
<head>
<script src="https://canvasjs.com/assets/script/canvasjs.min.
js"></script>
<script>
window.onload = function () {
var chart = new CanvasJS.Chart("svg", {
    animationEnabled: true,
    exportEnabled: true,
    theme: "light1",
    title:{
        text: "Test pyramid"
    },
    data: [{
        type: "pyramid",
        yValueFormatString: "#\"%\"",
        indexLabelFontColor: "black",
        indexLabelFontSize: 16,
        indexLabel: "{label} - {y}",
        dataPoints: [
            { y: 100, label: "Unit tests" },
            { y: 65, label: "Integration tests" },
            { y: 45, label: "UI tests" }
        ]
    }]
});
chart.render();
}
```

```
</script>
</head>
<body>
<div id="svg"></div>
</body>
</html>
```

This chart looks like this:

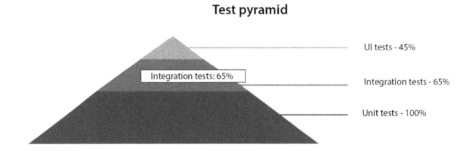

Figure 7.5: Pyramidal chart result

Note that we could load a `.csv` file containing these values, as well as the ones shown in the previous example.

Another interesting graph that we might want to create is one that visualizes the application crawler. Let us get started:

1.  First, let us define the JSON with the data. We are simplifying this here, but we could generate it with the crawler:

### CrawledData.json

```
{"nodes":[{"id":":"MainPage","group":1},{"id":"CartCheck-
out","group":1},

{"id":"ListItems","group":1},{"id":"MoreIn-
fo","group":1}],

"links":[{""source":"MainPage","target":"CartCheck-
out","value":1},
```

```
     {"source":"MainPage","target":"ListItems","value":8},
     {"source":"ListItems","target":"MoreInfo","value":10}]}
```

2. Now, let us see the HTML document:

## GraphExample.html

```html
<!DOCTYPE html>
<meta charset="UTF-8"/>
<html>
<head>
<script type = "text/javascript" src = "https://d3js.org/
d3.v4.js"></script>
</head>
<body>
<div><br></div>
<chart></chart>
<script src="graphExample.js"></script>
</body>
</html>
```

Finally, we need the JavaScript that contains the code logic for creating the chart, as shown in the following piece of code. The code for adding the labels to the text in the chart was created thanks to the hints provided in *[4]*. First, let us learn how to initialize the chart and its dimensions:

## graphExample.js – initializing the chart

```javascript
d3.json("crawledData.json", function(d) {
    createChart(d);
});
function createChart(data) {
    // set the dimensions and margins of the chart
var margin = {top: 10, right: 30, bottom: 30,
                left: 40},
  width = 400 - margin.left - margin.right,
  height = 400 - margin.top - margin.bottom;
```

3.    Then, we must select the chart object from the HTML body and add the graphics (svg):

## graphExample.js – selecting the chart and adding svg

```javascript
// append the svg object to the body of the page
var svg = d3.select("chart")
.append("svg")
  .attr("width", width + margin.left + margin.right)
  .attr("height", height + margin.top + margin.bottom)
.append("g")
  .attr("transform",
        "translate(" + margin.left + "," + margin.top +
")");
```

4.    Now, let us add the connections:

## graphExample.js – adding connections

```javascript
var link = svg.append("g")
    .selectAll("line")
    .data(data.links)
    .enter()
    .append("line")
    .style("stroke", "#777")
```

5.    Next, we must add the nodes and circles to the chart. We add nodes and circles instead of just circles so that we can add the labels (text) to the same object as the circles. This ensures they are visualized on top of the circles:

## graphExample.js - adding nodes and circles

```javascript
var node = svg.append("g")
  .selectAll("g")
  .data(data.nodes)
  .enter().append("g")
var circles = node.append("circle")
  .attr("r", 45)
    .style("fill", "#777777");
```

6.  In the same fashion, we must add the labels (text):

## graphExample.js - adding labels

```
var lables = node.append("text")
       .text(function(d) {
         return d.id;
       })
       .style("font-size", 5)
       .style("text-anchor", "middle")
node.append("title")
       .text(function(d) { return d.id; });
```

7.  Finally, we must add the simulation to position everything together, with the nodes repelling each other and staying centric (that is, to adjust them to the length of the object). The `ticked` function is used to readjust the position of every element in the simulation, specifying how nodes and links relate:

## graphExample.js - adding simulation

```
var simulation = d3.forceSimulation()
    .force("link", d3.forceLink()
           .id(function(d) { return d.id; })
    )
    .force("charge", d3.forceManyBody().strength(
      -2500))
    .force("center", d3.forceCenter(width / 2,
                                    height / 2))
    simulation
    .nodes(data.nodes)
    .on("end", ticked);
simulation.force("link")
       .links(data.links);
function ticked() {
  link
       .attr("x1", function(d) { return d.source.x; }
           )
       .attr("y1", function(d) { return d.source.y; }
```

```
            )
        .attr("x2", function(d) { return d.target.x; }
            )
        .attr("y2", function(d) { return d.target.y; }
            );
    node
        .attr("transform", function(d) {
          return "translate(" + d.x + "," + d.y + ")";
        })
    }
  };
```

The result of the preceding code looks like this:

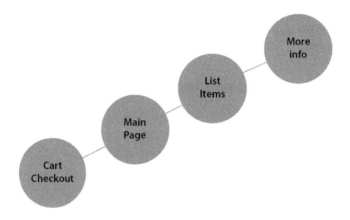

Figure 7.6: Result of the GraphExample.html file

You could add interactivity and other cool stuff to these graphs; the ones we've looked at only showcase what testing-related information or data you could display or need to display.

In this section, we learned how to easily create some charts and display the data related to our quality.

In the next section, we will look at an application that gives priority to test cases in a scientific way, rather than by using instinct. It would be impossible to do such an analysis if we hadn't gathered any data previously.

# Reviewing some techniques for test case analysis

If you were to ask different experts to categorize test cases by their priority, you would likely come across very different opinions. This task is not trivial, as having the right categorization could help us identify which test cases to automate first, or which ones to execute on our **build verification tests** (**BVTs**) so that we could promptly find the maximum number of issues and identify the most important tests cases that would cover them.

Most of the audience members I have asked throughout several conferences affirm to have more than 100 and even up to 500 test cases to explore per deployment or to automate. Handling so many test cases per deployment is currently achievable thanks to cloud parallel testing platforms (more on that will be covered in *Chapter 9, Having Your Head up in the Clouds*). However, these tools might charge us money per run or per the number of tests executed, and we must ensure those tests are bringing value to the deployment.

I have frequently found test suites containing way over 1,000 test cases, with only 10-20% of them actually bringing most value. Executing all of them exhaustively could bring up costs and time. Manually deciding which ones to execute, cleaning them, or automating them is also a costly manual effort.

One important attribute for someone working on a company's quality is to be able to balance the testing appropriately. Too little testing could incur in a poorly written application, which could look bad to the users and produce costs to fix the issues, including adding context switches. Too much testing might be counter-indicative and incur monetary or timely costs related to the creation, execution, and maintenance of those tests.

Here are some techniques we can use to analyze test cases to enhance the quality of delivery speed and the quality of the feature:

- Keep a clean test suite:

  - Decide which tests to create, maintain, remove, and execute per deployment.

  - Ensure priorities are updated if need be

  - Remove obsolete tests

  - Beware of tests that never passed or never failed

  - If it becomes too complicated, consider starting from 0

- Limit test repetitions across the test pyramid

- Keep good testing documentation to make sure everybody has similar concepts on priority, risks, etc.

- Periodically review the priority of the tests

- Keep BVTs alongside the latest extensive testing and discovery testing

- Perform extensive testing in a different environment before delivering to the user (use canaries, and as we explained in *Chapter 6, Continuous Testing – CI/CD and Other DevOps Concepts You Should Know,* for example, or have beta users that understand they may find issues with the app) and have the tests execute in parallel when possible

- Keep the usage hours of the application in mind and avoid peaks of activity

### What should go in the BVTs?

One of the goals of test case analysis is to decide which test cases are going to be used to verify the builds. The common agreement here is that there should be a small number of tests, just good enough to verify that nothing important in an application is broken. Some experts add that they should run in under 10 minutes, but I feel that this is a bit arbitrary and each application is different. Using a percentage of the total tests would have the same issue. Running all cases with the highest priority could be problematic if the priority is not well-taken care of and agreed upon.

When building a BVT suite, make sure you cover anything risky in your application in terms of damage to the company or the user that is monetary, physical, or otherwise. The main functionality of the application should also be checked. Keep a good balance and do not test too much either; try to cover as much as possible in as little time as possible.

While we could benefit from automatically analyzing the test cases, we generally need to include an expert to perform this task, which means that it is rather subjective and also hard to automate. However, you have read the name of this book – this is precisely the sort of task that we like and welcome here! In the next section, we will cover some early concepts in AI and try to find a mathematical solution to this automation.

## The early approach to AI – it's all thanks to mathematics

Artificial intelligence is, at its core, a set of mathematical functions and algorithms that help us achieve a purpose. Different purposes require different algorithms. Therefore, if you want to learn and/or use artificial intelligence to help you engage the quality of your system, it would be convenient for you to learn to love mathematics and learn some core mathematical concepts.

Lucky for us, some experts dedicate themselves to perfecting these algorithms so that we can use some predefined functions rather than having to write them on our own. However, as we discussed at the beginning of this chapter, if you know what the functions do, you will understand what to look for and which one of them to use.

There are different sets of definitions for AI. My favorite one is that of Luger and Stubblefield in 1993:

> **Artificial intelligence**
> *"[AI is the] branch of computer science that is concerned with the automation of intelligent behavior."*

This definition embraces the concept of automation, which is one of the specialties of test engineers (specially SDETs), so we could say that AI is intrinsically correlated with the testing discipline. In this case, we will not be automating a manual test but an intelligent behavior – the test analysis.

Let us revisit the way we have been approaching automation throughout this book; we will approach this task in the same manner:

1.  Identify repetitive tasks.
2.  Write code that would do these tasks for you.
3.  Identify when the code needs to be executed.
4.  Identify success measures.

## Identify repetitive tasks

Identifying repetitive tasks might sound like a trivial pursuit, but for tasks that come to you naturally and automatically, it might be harder than it seems. In cases such as this, we could use the technique of the "5 whys" to discover the exact task that's taken place. However, this time, we would use "5 hows" instead:

- We want a good balance in our tests. How?
- We analyze the tests and make decisions on which ones to create, maintain, remove, or execute. How?
- We order the tests. How?
- We make up a series of rules on them to bring them up or down the order. How?
- We mentally identify a set of variables and assign values to them to understand the priority of a test compared to others.

Now, we have a good idea of what exactly are our brains doing when we try to balance tests, so we can start the automation in the reverse order.

Examples of these variables (you can probably think of some more) are the *date of creation of the test, date of last execution of the test, manually assigned priority, risk or severity of failure of the test, number of bugs found by the test before, how long it takes to execute/automate the test, production usage*, and others.

Note that the variables that we take into account might not be always recorded alongside the test. It seems to me that people tend to spend less and less time recording them to speed up the process. Unfortunately, when it comes to anything related to AI, the more data we have, the better. Therefore, make sure you record the variables you need.

One extra detail to notice here is that some of the variables listed are subjective to the person that recorded them, such is the case of the priority of the testcase. These variables make the automation process a little bit more complicated, but there are ways to help with them, as we will see shortly.

Let us move to the second task regarding automation.

## Write code that would do these tasks for you

There are different approaches to this step and the one we'll discuss here may not be the best for your particular application. As mentioned previously, we want to inspire you and give you tips to achieve a better system so that you can apply the ones more convenient to your particular case or come up with new and better ones.

According to what we have mentioned in our "5 hows," the next step would be to automate the points assignment based on the variables. You could try to come up with an automated system to do this. For instance, if the test case was never passing or never failing, we could subtract a point. However, if the test case is new or has been manually assigned a high priority, or is for a risky feature (such as those involving payment), we could add a point. Then, we could sort the test cases by the points and pick those with more points.

Creating this system would be quite tedious to program, prone to errors, and likely hard to test. Besides, there are variables where we do not have a clear value assigned or a value we cannot trust.

This way of adding and removing weights is the core concept of a rule-based system.

> **Rule-based system**
> This is a system in which we store and manipulate data to give it a useful interpretation.

In this system, we would define a series of rules, writing them in human-readable language and a logical form. This group of rules is called a **knowledge base**.

These rules will not be automatically generated (in this case; we will see more on this in the next chapter), but human-created. That means we still need an expert or experts to generate the rules, but then anybody can use them.

As we mentioned previously, some variables have a clear value (for example, the age of a test case), while others have a less clear value (for example, what is a "high" priority?). Instead of indicating points directly, we will be using functions so that we can estimate less precise values.

> **Fuzzy logic**
> **Fuzzy logic** is a technique that's utilized to model logical reasoning with vague or imprecise statements.

In a fuzzy system, we define the values of the variables using percentages, probabilities, and functions rather than fixed numbers. An expert could guess the function that a variable would use, although it would be better to try different functions in the system and validate that they behave correctly. We will learn how to do that at the end of this section.

The typical functions that we will find are trapezoid (*Figure 7.7*), triangle (*Figure 7.8*), and Gaussian (*Figure 7.9*), although there are others too:

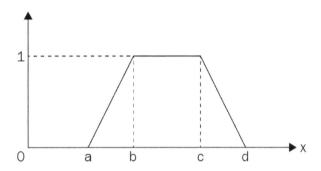

Figure 7.7: Trapezoid function

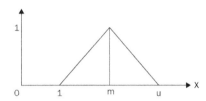

Figure 7.8: Triangle function

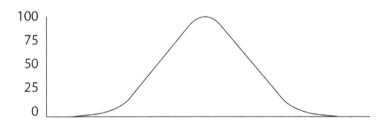

Figure 7.9: Gaussian function

Once the functions have been defined, the rule-based system's algorithm will go through some cycles to make a final decision. In our case, the decision would be the importance of the test case (for automation or execution) compared to the other test cases.

The way the algorithm works to decide which rule to apply and what percentage of certainty is as follows:

- **Match**: There would be a first pass trying to match all the possible rules with each test case while considering the variables associated with them. If there are conflicts (which is common), they are pushed to a conflict set.

- **Conflict resolution**: There is another pass that checks the rules that conflict with each other and tries to select the most likely rule to apply from among them. If the conflicts cannot be resolved or no rule can be selected, the system will fail. Therefore, we must have the right rules to cover all the cases.

- **Act**: Finally, the system marks the test cases with the likely resolution label (for example, automate or not automate with a confidence of 80%).

## *A practical example of a rule-based system for testing*

In this section, we are going to look at a practical example of the use of a rule-based system that will help us decide which test cases to execute or automate from a given list, using the details specified in the rules and the variables that we have recorded for those test cases.

The best-suited programming languages to create a rule-based system are those prepared to execute logical programming, such as R, F#, and others. However, we could use any other programming language, so long as it provides a set of libraries to assist us.

A programming language that is commonly used for AI that has a lot of AI-related libraries is Python. We will look at some Python AI-related libraries in the next chapter, but for comparison purposes, we will not be using them in this section. If you are interested in implementing this using Python, look at the **skfuzzy** *[5]* library.

For this practical example, we will be using C# with a library for AI-related functionality called **accord** *[6]*. You should be able to install its NuGet package with Visual Studio (or an equivalent library).

We are going to define our system's items as follows:

- **Antecedents or inputs**: These are the variables that we will take into account in our rules. In our system, these are the following:

  - Priority

  - Number of steps

  - Age of the test case

  - Pass-failure rate

- **Control system**: The rules we (an expert) have defined.
- **Consequent or output**: The result variable:

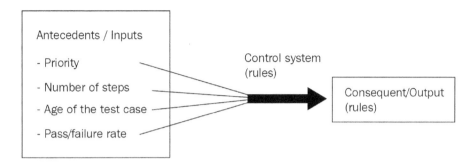

Figure 7.10: Visualization of the rule-based system

- **Priority**: We will define three categories for priority and their functions, based on the certainty percentages in which we could say that a priority lands on each of the categories:

  - **Low** would be a trapezoidal function starting on the edge (minimum value or 0) and traveling between 20% and 40% certainty

  - **Medium** would be the same type of function on the 20%, 40%, 60%, and 80% certainty (note that they overlap)

  - **High** can be defined between 60%, 80%, and the maximum value (in this case, 100%)

The graph would represent our system as such:

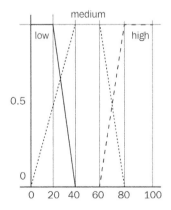

Figure 7.11: Priority visualized

- **Steps**: Similarly, we will define the number of steps with another set of trapezoidal functions (see *Figure 7.12*):

  - **Low** going from minimum (0) to 6

  - **Medium** going from 5 to 9, to 12, and finally to 15

  - **High** going from 10 to the maximum number of steps, which in our case is 15:

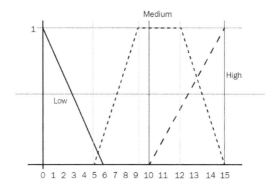

Figure 7.12: Steps visualized

- **Age**: The age of the function will also be represented in three sections, each of them with a trapezoidal function (see *Figure 7.13*):

  - **New**: 0 – 30 days old

  - **Medium**: 15 – 50 – 70 – 90 days old

  - **Old**: 80 – 100 days old:

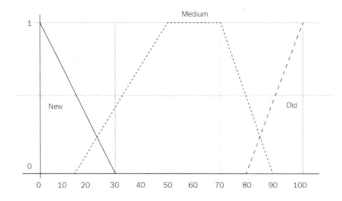

Figure 7.13: Age visualization

- **Pass rate**: Before assigning the values of this function, we measure the number of maximum executions of the test case (passes and failures) and divide this number by the number of passes. That gives us a percentage that we could use in the function from 0 to 1. This time, the function to use is not going to be trapezoidal but triangular, stating the certainty of whether something is passing (yes) or failing (no) (see *Figure 7.14*). Note that we could have also used a Gaussian function as it might produce better results. You should tailor this code and find the best function for your system (see *Figure 7.15*):

  - **Yes**: 0 – 60

  - **No**: 40 – 100:

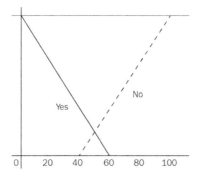

Figure 7.14: Pass rate visualization

- Lastly, we need to add the values for the consequent (if something should execute, or automate, which could be marked as yes or no). Again, we could use a triangular function (for simplification), but in our empirical tests, we found that a Gaussian function works best. If you are considering this system, make sure you test this option too. The graph for the triangular function would be as before. The graph for the Gaussian function would be as follows:

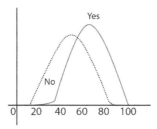

Figure 7.15: Gaussian function example

If we decide on the variables (for example, "priority") and definitions (also called labels – for example, "low"), the functions that compose those labels (as shown in the preceding graph), and the rules among the variables, we should be able to implement a system that will decide for us whether we should run a test or whether it is safe to go without it. Let us do so in the following example part of the `ruleBasedSystem.cs` file:

1. The first thing we need to do is define a fuzzy database where we will keep all these definitions:

```
Database fdb = new Database();
```

2. Then, we need to create linguistic variables representing the variables we want to use in our system. In our case, we want to look at the following:

```
LinguisticVariable priority = new
LinguisticVariable("Priority", 0, 100);
LinguisticVariable steps = new LinguisticVariable("Steps",
0, 100);
LinguisticVariable isNew = new LinguisticVariable("IsNew",
0, 100);
LinguisticVariable isPassing = new
LinguisticVariable("IsPassing", 0, 100);
```

3. Finally, we want to define a linguistic variable where we will store the result. We will call this `MarkExecute`:

```
LinguisticVariable shouldExecute = new
LinguisticVariable("MarkExecute", 0, 100);
```

4. After that, we must define the linguistic labels (fuzzy sets) that compose the preceding variables. For that, we need to define their functions, as indicated previously. Let us start with the ones related to priority:

```
TrapezoidalFunction priorityLowFunction = new
TrapezoidalFunction(20, 40, TrapezoidalFunction.EdgeType.
Right);
FuzzySet priorityLow = new FuzzySet("Low",
priorityLowFunction);
TrapezoidalFunction priorityLowFunction = new
TrapezoidalFunction(20, 40, 60, 80);
FuzzySet priorityMedium = new FuzzySet("Medium",
priorityMedium);
TrapezoidalFunction priorityHighFunction = new
TrapezoidalFunction(60, 80, TrapezoidalFunction.EdgeType.
Left);
```

```
FuzzySet priorityHigh = new FuzzySet("High",
priorityHighFunction);
```

5.  Now, we can reuse the same functions for other values, or even just add the same labels for simplicity. However, in our system, we want to add slightly different functions, so we will add new ones for each variable:

```
TrapezoidalFunction stepsLowFunction = new
TrapezoidalFunction(0, 6, TrapezoidalFunction.EdgeType.
Right);
FuzzySet stepsLow = new FuzzySet("Low",
stepsLowFunction);
TrapezoidalFunction stepsMediumFunction = new
TrapezoidalFunction(5, 9, 12, 15);
FuzzySet stepsMedium = new FuzzySet("Medium",
stepsMediumFunction);
TrapezoidalFunction stepsHighFunction = new
TrapezoidalFunction(10, 15, TrapezoidalFunction.EdgeType.
Left);
FuzzySet stepsHigh = new FuzzySet("High",
stepsHighFunction);
```

6.  Now, let us define the triangular functions. For accord-framework, we cannot use no as it is a *reserved word*, so we are calling it Nope

    I.  For the age variable, we must define the functions as shown in *Figure 7.13*:

```
TrapezoidalFunction oldFunction = new
TrapezoidalFunction(0, 30, TrapezoidalFunction.EdgeType.
Right);
FuzzySet ageOld = new FuzzySet("Old", oldFunction);
TrapezoidalFunction ageMediumFunction = new
TrapezoidalFunction(15, 50, 70, 90);
FuzzySet ageMedium = new FuzzySet("Medium",
ageMediumFunction);
TrapezoidalFunction ageNewFunction  = new
TrapezoidalFunction(90, 100, TrapezoidalFunction.
EdgeType.Left);
FuzzySet ageNew = new FuzzySet("New", ageNewFunction);
```

II.   For the passing variable, we must define the functions as shown in *Figure 7.14*:

```
TrapezoidalFunction notPassFunction = new
TrapezoidalFunction(10, 50, TrapezoidalFunction.EdgeType.
Right);
FuzzySet notPassing = new FuzzySet("Nope",
notPassFunction);
TrapezoidalFunction passingFunction = new
TrapezoidalFunction(50, 90, TrapezoidalFunction.EdgeType.
Left);
FuzzySet passing = new FuzzySet("Yes", passingFunction);
```

III.  For the consequent variable (`should execute`), we could reuse the previous function or add two new functions according to *Figure 7.14*, as shown in the following code:

```
TrapezoidalFunction doNotExecuteFunction = new
TrapezoidalFunction(10, 50, TrapezoidalFunction.EdgeType.
Right);
FuzzySet doNotExecute= new FuzzySet("Nope",
doNotExecuteFunction);
TrapezoidalFunction executeFunction = new
TrapezoidalFunction(50, 90, TrapezoidalFunction.EdgeType.
Left);
FuzzySet execute = new FuzzySet("Yes", executeFunction);
```

IV.   Then, we must add these labels to the variables:

```
priority.AddLabel(priorityLow);
priority.AddLabel(priorityMedium);
priority.AddLabel(priorityHigh);
steps.AddLabel(stepsLow);
steps.AddLabel(stepsMedium);
steps.AddLabel(stepsHigh);
isNew.AddLabel(ageNew);
isNew.AddLabel(ageMedium);
isNew.AddLabel(ageOld);
isPassing.AddLabel(passing);
isPassing.AddLabel(notPassing);
shouldExecute.AddLabel(execute);
shouldExecute.AddLabel(doNotExecute);
```

V.  Finally, we must add the variables with the labels to the fuzzy database that we defined at the beginning:

```
fdb.AddVariable(priority);
fdb.AddVariable(steps);
fdb.AddVariable(isNew);
fdb.AddVariable(isPassing);
fdb.AddVariable(shouldExecute);
```

Now that we have defined the system, let us create the rules.

7.  For `accord.net`, we should create an inference system so that we can assign some rules to it:

```
InferenceSystem IS = new InferenceSystem(fdb, new
CentroidDefuzzifier(1000));
```

The rules are defined in plain English so that it is easier for the experts and other players in the development to review them and agree upon them. In our case, we are not giving any different weights to the rules; they will all be "as important" but we could add them if we find some are more necessary than others. In the following code, we have defined eight rules, but your system may need more or would work better with more:

```
IS.NewRule("Rule 1", "IF Priority IS High THEN
MarkExecute IS Yes");
IS.NewRule("Rule 2", "IF Priority IS Medium AND IsPassing
IS Nope THEN MarkExecute IS Yes");
IS.NewRule("Rule 3", "IF Priority IS Medium AND IsNew IS
Medium then MarkExecute IS Yes");
IS.NewRule("Rule 4", "IF Priority IS Low AND IsPassing IS
Yes THEN MarkExecute IS Nope");
IS.NewRule("Rule 5", "IF Priority IS Low AND IsNew IS Old
THEN MarkExecute IS Nope");
IS.NewRule("Rule 6", "IF Steps IS High AND IsPassing IS
Nope THEN MarkExecute IS Yes");
IS.NewRule("Rule 7", "IF Steps IS Low AND Priority IS
Medium THEN MarkExecute IS Yes");
IS.NewRule("Rule 8", "IF Steps IS High AND Priority IS
Medium THEN MarkExecute IS Nope");
```

8.  Finally, we need to set the actual inputs (antecedents) or specific values from the test cases. As we saw in *Chapter 3, The Secret Passages of the Test Pyramid – The Middle of the Pyramid*, we could use our test case database's API to automate how these variables are extracted from our tests into a `.csv` file.

9.  As a quick test, we could type the values directly, as shown in the following code. For example, let us think of a new test case (3 days ago) that has a low number of steps (3), a low priority (20%), and a low passing rate (since it is new) of 10%. This would look like this:

```
IS.SetInput("IsNew", 3);
IS.SetInput("Steps", 3);
IS.SetInput("Priority", 20);
IS.SetInput("IsPassing", 10);
```

10. To define an old test case (90 days) with a high priority (90%), a high number of steps (20), and with high passing rate, we could use the following variables:

```
IS.SetInput("IsNew", 90);
IS.SetInput("Priority", 90);
IS.SetInput("Steps", 20);
IS.SetInput("IsPassing", 90);
```

Ideally, we would like to analyze all the test cases in our system, produce a file as a result of executing this code, and have all the tests in there organized by their certainty of execution.

11. For simplicity, let us retrieve the consequent or result variable and print it in the console:

```
try
{
float testCaseResult = IS.Evaluate("MarkExecute");
Console.WriteLine("The test case should be " +
testCaseResult + " executed.");
Console.ReadKey();
}
catch (Exception e)
{
Console.WriteLine("We found an exception : " +
e.Message);
Console.ReadKey();
}
```

In this section, we looked at a practical example of how to identify test cases that should be executed or not, which could help us identify BVTs and re-prioritize test cases.

## Improving the system

This system might not be ideal for your application. Many factors could make it better; for example:

- Try using different functions for the different variables of the antecedents.

- Add more variables or use different ones.

- Try changing the rules or adding new ones.

- Change the consequent minimum threshold. In our case, we have it set to 50%, but if the system allows too many tests to execute, we could use a higher value or vice versa. It is a similar idea to using different functions or different peaks for the functions for the variables.

We will also need to be able to estimate how well the system is performing. To do so, we could save the last result from all the cases and take it out of the total. Then, we could compare how well the prediction is doing against the last actual run.

Finally, we should iterate through all of the tests and check the result percentage so that we can sort by that and get the number of steps:

```
try {
    String[] lines = File.ReadAllLines(@"testcases.csv");
    StringBuilder testResults = new StringBuilder();
    foreach (String line in lines) {
        String[] test = line.Split(',');
        IS.SetInput("Priority", float.Parse(test[0]));
        IS.SetInput("Steps", float.Parse(test[1]));
        IS.SetInput("IsNew", float.Parse(test[2]));
        IS.SetInput("IsPassing", float.Parse(test[3]));
        float testCaseResult = IS.Evaluate("MarkExecute");
        testResults.Append(line).Append(',').Append(
            testCaseResult);
        testResults.Append(';').AppendLine();
    }
    File.WriteAllText("testResults.csv",
                      testResults.ToString());
}
catch (Exception e) {
    Console.WriteLine("Exception found: " + e.Message);
    Console.ReadKey();
}
```

Now, we should be able to use any `.csv` visualizer to sort on the last column and inspect the data, or we could try that bit programmatically.

## Identify when the code needs to be executed

Now that our code is up and running, we should decide when we need to execute it. Since this code could have different applications, this would depend on what we use it for:

- Finding incorrect/flaky test cases
- Finding best value tests for a BVT or even an interactive BVT (which would change with new test cases)
- Reducing cost and time in test runs
- Making quick validations under a specific, set time or number of steps
- Increasing the understanding of the old system and avoiding rewriting tests

For instance, if we want to create a set of BVTs, we should execute this over integration tests (this is preferable if they have been already executed a few times, so we could add the passing rate). When we include a new set of tests, we should re-execute this over the entire system to decide which ones to pick. That would also be convenient for removing old test cases.

If there are particularly risky tests, you could opt for adding a particular variable for them (for example, risk) or you could opt for taking them out of the system and ensuring you always have those in the BVTs.

## Identify success measures

When working with AI, the system already gives us success measures. Besides this, we will be able to see how the system behaves executing the testcases selected in respect with previous executions.

In order to improve over iterations, we could keep repeating the automation process. We could add another layer of automation on top of this, aligning it with the CI/CD structure we talked about in the previous chapter so that every time new integration tests are added to the system, the rules get executed and a new set of BVTs is created.

We should do this if we want to use this to create a set of BVTs, but if we wish to use it to decide which test cases automate earlier, then executing it after writing the definition of the (manual) test cases would be enough, without the need to automate this execution.

If you already have a lot of test cases written and would like to cover a group of different ones but not all of them, consider a random test crawler.

Creating a random test crawler would be quite trivial and there would be many different ways of doing so, depending on your system. Therefore, we have not included the code for that here, but we encourage you to try creating one. For instance, remember the code that we used to execute tests with **vstest** in the previous chapter; check the parameters that the tool allows and think about how you could just randomly call the tests.

> **Remember**
> Tests should be completely independent of each other and leave the system in the same state as it was before each of the particular tests were executed.

In this section, we looked at some techniques we can use to decide between test cases, and how mathematics can help us improve them and find solutions using artificial intelligence.

In the next chapter, we will look at another way of dealing with this automation by using a full artificial intelligence solution, introducing machine learning.

## Summary

We are more than halfway through this book – well done on getting here! If you are reading this, chances are you are enjoying this book. If that is the case, please consider telling others about it. Feedback is also very much welcome (so long as it is gentle; I am a kind and sensitive soul).

I hope this chapter has made you curious about mathematics and makes you want to keep an open eye for it everywhere and an open mind for it as well, enjoying the little magic inherent to it. In this chapter, we looked at several algorithms and concepts that can help you grow your mathematical skills. We also saw how to put some of them into practice with exploratory testing (creating a web crawler), showing our data by creating dashboards, and our first approach to artificial intelligence with the test cases analyzer.

So far, we have seen concepts that could be applied to most applications. However, they cannot be applied to all applications equally or straightforwardly. In the next chapter, we will cover tips and tricks for testing challenging applications and learn more about the name of this book. We will dive a bit deeper into the concepts needed to understand the basics of artificial intelligence and guide you on where you should go next if you are dealing with these sorts of systems.

## Further reading

For more information about the topics that were covered in this chapter, take a look at the following resources:

- [1] More about JavaScript graphical functions with the D3.js library: `https://d3js.org/`
- [2] Matplotlib can be found here: `https://matplotlib.org/`
- [3] The Canvas.js documentation for pyramid charts can be found at `https://canvasjs.com/javascript-charts/pyramid-chart/`

- [4] Another example of graph usage, along with information about adding labels, can be found at `https://gist.github.com/heybignick/3faf257bbbbc7743bb72310d03b86ee8`

- [5] Documentation on the scikit fuzzy library: `https://pythonhosted.org/scikit-fuzzy/api/skfuzzy.html`

- [6] Documentation on the accord framework: `http://accord-framework.net`

Make sure you review courses on artificial intelligence, fuzzy logic, algorithms, and all the items mentioned at the beginning of this chapter. A couple of resources from our library that might be interesting to follow up on are.

- `https://www.packtpub.com/product/master-math-by-coding-in-python-video/9781801074537`

- `https://www.packtpub.com/product/graph-machine-learning/9781800204492?_ga=2.189825387.158376204.1672672304-771539045.1672672304`

# Part 3
# Going to the Next Level – New Technologies and Inspiring Stories

In this part, we will travel one step further, with ideas about how to use new technologies, and we will get inspired to think out of the box and test challenging applications. These topics are not usually learned or applied to testing, but by applying them to your system, you can go beyond expectations.

This part comprises the following chapters:

# 8

# Artificial Intelligence is the New Intelligence

**Artificial intelligence** (**AI**) has a wide scope, and it can help with our testing in several ways. It can also help with creating automation tools. In the previous chapter, we provided an introduction to this topic and gave some examples of where AI could be useful. However, it would be impossible to cover AI fully in this book (consider that AI experts have dedicated their careers to the topic). In this chapter, we will provide an introduction to some of the core topics of AI and see examples of how it is (and could be) useful for testing. Then, we will review applications that use AI so that we can identify them and check whether we need to do anything special to test those. Finally, we will see some other technical examples that might be of use in your application.

In this chapter, we are going to cover the following main topics:

- Is AI taking my job? Why learn about AI?
- Core concepts of AI
- AI for testing
- Testing AI apps
- Sending your code to the Gym
- Other cool AI projects and tools

## Technical requirements

This is an advanced chapter and while we hope you can still enjoy it without developing mathematical/ AI knowledge, you would get the best out of this chapter if you are familiar with these topics and have some degree of programming knowledge. However, you can still take this one as an introduction so that you can learn more on your own. We hope to make it as fun and easy as possible so that it can help you build useful tools for your team!

The examples in this chapter were written using Python, although they could be written in different languages. You can find them in this book's GitHub repository: `https://github.com/PacktPublishing/How-to-Test-a-Time-Machine/tree/main/Chapter08`.

# Is AI taking my job? Why learn about AI?

In the past, I've given talks about AI topics in which I could see faces of awe and enjoyment but also fear among the audience. The most frequent questions from the audience with little AI knowledge were the ones posed in the title of this section:

- Is AI going to replace the job of a tester/QA expert?
- Why should I learn about AI?

More experienced people who are into testing (and even any other job, especially ones related to computing) also wonder and frequently argue about the first question.

To answer the first question, we should first define and agree on what exactly AI is. In *Chapter 7, Mathematics and Algorithms in Testing*, we saw the following AI definition:

> **Artificial intelligence – Luger and Stubblefield, 1993**
>
> *"[AI is the] branch of computer science that is concerned with the automation of intelligent behavior."*

From this definition, we can easily conclude that AI is a tool for automation. Therefore, the question of *AI taking over the testing job* could be rephrased as *automation taking over the testing job*.

## Will automation take my job?

From the trends we are seeing in testing and automation, we can see that automation is taking a more serious role versus manual testing, but positions in tests are still needed. People in these positions are becoming specialized in writing automated test cases and learning automation. However, there are other people (developers) that are comfortable creating code and are starting to get their share of automation writing (shifting left). There will always be the need for people with test expertise, but as with other areas of computer science, renewing, growing, and learning are vital to keeping up with the trends and with the natural development pace of computer science.

In my opinion, what will happen with AI is similar: there will be experts first that are in charge of writing the tools and who could be consulted for expertise, but people in testing jobs will start learning the basics for them to create tools as well and to tailor the testing accordingly. People grow and job descriptions change a little, but the expertise for testing will still be required. Even if development could be taken by AI, tools will still need some degree of quality checks and verifications, even if to make sure that the data provided to the algorithms is correct and ethical.

## Why should I learn about AI?

The previous logic also brings us to a conclusion for the second question: learning AI will help you grow and build unique, specialized tools that your organization could use for testing.

Even if you do not believe this would be a trend, consider the advantages of learning just a little bit of AI. As we mentioned previously, this could be of help to automate applications and write tools that could be very difficult to write otherwise. In general terms, you should consider writing an AI application if any of the following apply:

- There are no current algorithms to solve the problem or they are hard to create

- The existing algorithms are too complex

- The existing algorithms have bad results

- There could be a relationship between data that a human might not naturally think of

Now that we have argued the importance of learning AI in the testing spectrum, let us review some core concepts of AI to get us started with it and use it for creating tools for testing.

# Core concepts of AI

In the previous section, we saw one of the AI definitions that relates AI to automation. However, the most commonly used definition for AI is that of Arthur Samuel:

> **Artificial intelligence – Arthur Samuel, 1959**
>
> AI gives "*computers the ability to learn without being explicitly programmed.*"

This definition, while being more general about the "what," is more specific and restrictive about the "how" than the one mentioned in the previous section. It sets the goal of AI to *learn on its own* rather than *appear smart* as we could make it seem in the previous chapter by programming rules. This *apparently* subtle difference is the common base for debate in the scientific community and it is likely to get more and more restrictive as the techniques and machines advance toward the goal of intelligence per se.

There are different areas in AI. To continue with the topic of machine learning, let us define it:

> **Machine learning – IBM's definition**
>
> "*A branch of AI focused on building applications that learn from data and improve their accuracy over time without being programmed to do so.*"

The counterpart of using machine learning versus other AI systems is that it requires a lot of data to train the system. Inside the machine learning area, there are different approaches to learning.

When creating systems for learning for a machine, we tend to look at intelligent systems that are currently in the world. For example, we could look at how the human brain works and try to simulate it. That is the case with *neural networks*, and, more specifically, **deep learning**:

> **Deep learning**
>
> Deep learning is a part of AI that is based on a leveled artificial neural network structure, simulating that of the human brain.

Another subfield of AI that is gaining momentum, with tools such as **GitHub Copilot** *[1]*, or chatGPT *[2]*, and that we might want to start looking for applications in the testing area is **natural language processing** (**NLP**), which emulates human language understanding:

> **Natural language processing**
>
> Natural language processing is a part of AI that focuses on making the computer capable of understanding natural human language.

As well as understanding processed language (generally in the form of text), machines sometimes need to be able to understand things visually, which takes us to the term **computer vision** (**CV**):

> **Computer vision – Huang, T. (1996-11-19)**
>
> *"[...]From an engineering point of view, computer vision aims to build autonomous systems that could perform some of the tasks that the human visual system can perform (and even surpass it in many cases)."*

In other words, computer vision is a part of AI that enables computers and systems to analyze videos, digital images, and other visual inputs and derive useful information from them, to then reach intelligent conclusions.

An example of a CV system would be the case of an application that automatically tags images based on their contents. We will review some of these techniques in due time, although we recommend that if you are not already familiar with them, you seek to learn about them in more depth. In the next section, we will look at different ways a machine could learn.

## Types of learning

We could distribute the AI applications in groups depending on the techniques used to solve the problem and the type of problem we are trying to solve. This is important for us because different applications should be tested in different ways. We'll look at these types of learning in this section and keep them in mind when we talk about AI applications later in this chapter.

## Supervised learning

In supervised learning, we map labeled inputs to outputs in an offline model (that is, we train the system in an offline manner, then we can use the trained system online to match a real output).

We have a training set (that we will use for learning and therefore to train the system) and a dataset (which is the data that we want to evaluate).

This learning is used, for example, for classifying images or making predictions. An example of this is the one seen in the *Early approach to AI – it's all thanks to mathematics* section of the previous chapter. We will review this example in the *Data curation* and *Test case analysis with machine learning* sections while using a supervised machine learning system to solve it.

## Semi-supervised learning

In this system, some of the training examples are labeled and some are not. Therefore, only part of the training set could be applied to supervised learning.

There are different ways of approaching systems like this. A logical way to solve this is to use the labeled data for an initial training set; then, we could use an unsupervised system to label some of the unlabeled data and add it to the training set iteratively, based on the confidence of the data having the correct label. Finally, we could use the newly labeled data alongside the examples that were labeled from the beginning to perform supervised learning.

## Unsupervised learning

In unsupervised learning, we keep learning iteratively in an "online" way; we readjust the system with each iteration of learning and use each analysis to add to our training set, therefore, all data is both training data and real data. The system only receives inputs and uses real-time analysis to produce the outputs, which could be explicitly corrected if wrong, resulting in better predictions with each round.

These systems are the ones that generally scare people as the creators have less control over the outputs that the system would have and how it would evolve as it trains itself.

Examples of unsupervised learning could be clustering similar types by association (for example, for labeling data in the semi-supervised learning example).

## Reinforcement learning

In reinforcement learning, we add some sort of mathematical reward or punishment to each iteration of the code. There is no need for labels for input and outputs, prepared datasets, or actions explicitly being corrected. Instead, the system looks dynamically at the current state versus what it has left to explore. These systems are very suitable, among other applications, for finding goals, which can help us to automatically win games, which are generally hard to automate.

The entity that performs the reinforcement learning is known as an **intelligent agent**:

> **Intelligent agent**
>
> An intelligent agent is an autonomous entity that acts on an environment to try to achieve goals and maximize a reward by what it observes in such an environment.

Therefore, reinforcement learning can be considered an area of machine learning that uses agents.

There are two important concepts related to reinforcement learning that might be confusing. Let us define them here for clarification:

- **Exploration**: The agent explores to find information about the environment
- **Exploitation**: Using the information found during the exploration, the agent tries to get the maximum reward

Another important term is **episode**, which is the maximum number of iterations or states the agent can use to try to find a solution in the environment.

Before we perform any of these learnings, we must measure and provide enough precise data in the right way for the machine to use as a learning point. Therefore, we feel we must have a section explaining how to clean and extract data in depth.

## Data curation

When we use variables for AI, it is of vital importance that they are properly curated and normalized so that the results are gathered correctly and the machine can proceed to accurate learning. We should also try to use the same forms of measurement across all of them (for example, if you use seconds, multiply minutes by 60).

There are various ways of doing this. Let us see how to curate data by taking the example from the previous chapter, in which we analyzed test cases extracted from test rails into a `.csv` file. You can find such a `.csv` file here: `https://github.com/PacktPublishing/How-to-Test-a-Time-Machine/blob/main/Chapter08/testcases.csv`. So, let us get started:

1. Let us visualize the first few lines of such a file:

**testcases.csv**

```
id,fails,passes,Priority,CreatedDate,Steps,LastResult
100,1,9,3,2021-12-15,22,0
101,10,8,3,2021-08-11,13,0
102,1,5,3,2021-11-26,24,1
```

```
103,7,10,3,2021-11-05,27,0
104,8,10,2,2021-12-08,6,1
105,2,5,2,2021-11-16,2,1
106,2,2,3,2021-10-27,24,0
107,2,3,3,2021-11-12,11,0
. . . .
```

2. Now, we should make some changes to these variables so that they make sense in the system.

   For example, it would be more complicated for the system to extract intelligent information out of a date field than if we specify the number of days that have passed from the creation date to the current date. Then, we can divide them all by the oldest date to get a number that will be within a fixed range from *0-1*. We do a similar thing with the number of steps. Finally, multiplying by 100 will give us a percentage.

3. For failing and passing, we calculate the average of passes versus the total execution (passes and failures) and also multiply by 100.

4. We will use the **pandas** *[3]* library to load the data from a .csv file (don't forget to execute pip install for your Python libraries). Then, we will perform the aforementioned operations and, finally, save the new values into a new file.

5. In the following example, we are normalizing the variables that we had in the .csv file manually (please read the comments marked with # to understand what each part of the code is doing):

## dataCurator.py

```python
import pandas as pd
import datetime
# Retrieve the test cases from the csv into a
# dictionary
dataFile = open('testcases.csv', 'r')
dataFrame = pd.read_csv('testcases.csv', index_col=0)
dataDictionary = dataFrame.transpose().to_dict()
# Create a new dictionary to save the results
result = {}
# Data optimization: get the maximum number of steps
# and of differences
oldest = 0
maxSteps = 0
for t in dataDictionary:
```

```python
        createdDate = datetime.datetime.fromisoformat(
          dataDictionary[t]['CreatedDate'])
        currentDate = datetime.datetime.now()
        difference = currentDate - createdDate
        daysOfDifference = difference.days
        dataDictionary[t]['daysDiff'] = daysOfDifference
        if daysOfDifference > oldest:
            oldest = daysOfDifference
        if dataDictionary[t]['Steps'] > maxSteps:
            maxSteps = dataDictionary[t]['Steps']
    # Iterate again to save the data in the new dictionary
    for testcase in dataDictionary:
        result[t] = {}
        passes = dataDictionary[t]['passes']
        fails = dataDictionary[t]['fails']
        # Data optimization - get the percentage of
        # passes
        passingTest = (passes / (fails + passes)) *
                        100
        # Data optimization - get the percentage of
        # test age
        testAge = (dataDictionary[t]['daysDiff'] /
                  oldest) * 100
        priorityTest = dataDictionary[t]['Priority']
        # Data optimization - get the percentage of
        # number steps
        numSteps = (dataDictionary[t]['Steps'] /
                  maxSteps) * 100
        # Save the new values
        result[t]['Priority'] = float(priorityTest)
        result[t]['Age'] = float(testAge)
        result[t]['Steps'] = float(numSteps)
        result[t]['Passing'] = float(passingTest)
    # Save to the new Data File
    dataFrame = pd.DataFrame.from_dict(result,
    orient='index')
```

```
dataFile = open('newData.csv', 'w')
dataFrame.to_csv(dataFile, sep=',')
dataFile.close()
```

6. Let us review the first lines of the resulting file to see how the numbers are more suitable for a machine to learn (they are all percentages) but less obvious for a human to understand:

## newData.csv

```
,Priority,Age,Steps,Passing,Expected
100,3.0,10.526315789473683,73.33333333333333,90.0,0.0
101,3.0,93.42105263157895,43.333333333333336,44.444444444
44444,0.0
102,3.0,23.026315789473685,80.0,83.33333333333334,1.0
103,3.0,36.84210526315789,90.0,58.82352941176471,0.0
104,2.0,15.131578947368421,20.0,55.55555555555556,1.0
105,2.0,29.605263157894733,6.666666666666667,71.428571428
57143,1.0
```

We could normalize the data manually, as shown previously, or use some pre-built library to achieve the same purpose.

7. In Python, we could use `sklearn.preprocressing` for this, as shown in the following equivalent example (please read the comments marked with # to understand what each part of the code is doing):

## dataNormalizationSKLearn.py

```
# Data normalization
from sklearn.preprocessing import StandardScaler
import pandas as pd
import datetime
import time
# Retrieve the test cases from the csv
dataFile = open('testcases.csv', 'r')
dataFrame = pd.read_csv('testcases.csv', index_col=0)
sc = StandardScaler()
# Curating the data of the rest of the variables
# Note - passing should be input manually in this
# case, or automated as above / created date should be
```

```
# in timestamp
dataFrame['CreatedDate'] = [datetime.datetime.
fromisoformat(t).timestamp() for t in
dataFrame['CreatedDate']]
dataFrame['Passing'] =dataFrame['passes'] /
(dataFrame['passes'] + dataFrame['fails'] )
dataFrame[['CreatedDate', 'Steps', 'Passing']] =
sc.fit_transform(dataFrame [['CreatedDate', 'Steps',
'Passing']])
# Save to the new Data File
dataFile = open('newDataSKLearn.csv', 'w')
dataFrame.to_csv(dataFile, sep=',')
dataFile.close()
```

8.  Let us see the first few lines of the output file from executing the preceding code so that we can compare them with the manually created ones:

## newDataSKLearn.csv

```
id,fails,passes,Priority,CreatedDate,Steps,LastResult,
Passing
100,1,9,3,1.3812160296434677,0.7485500513501337,0,1.95255
2682920596
101,10,8,3,-1.4971565831863667,-0.2942603067604978,0,-
0.26837269321110785
102,1,5,3,0.947319199411929,0.9802856864858296,1,1.627539
2132427864
103,7,10,3,0.4677490186297021,1.3278891391893735,0,0.4326
367511920128
104,8,10,2,1.2213593027160587,-
1.1053350297354334,1,0.2733164229185764
105,2,5,2,0.7189524466584876,-
1.568806300006825,1,1.047158017389553
106,2,2,3,0.2612674130151321,0.9802856864858296,0,0.00247
1864853734247
```

Here, we can see floating-point numbers rather than percentages, so it is even less obvious for us to follow. However, for the machine, it is perfectly easy to calculate the weight of each of the labels for each of the test cases so that it can gather information and reach conclusions about how they relate.

9. Finally, we could do a mix of the two things, which could be handy as we had to calculate the percentage of passing tests. First, do some manual work to clear up the path and the obvious differences a little bit and finish with the use of some preprocessing library to make sure all the data is scaled in the same way. You can find the full code in the `dataNormalization.py` file on GitHub. The result of the output is very similar to the one shown previously; you can find it in the `newDataBoth.csv` file.

Now that we have some basic knowledge of AI, and we have normalized the data for our example, we will learn more about how to use machine learning as a tool for testing.

# AI for testing

In the previous chapter, we looked at an application of AI for testing: a test case analyzer. Later, in the *Test case analysis with machine learning* section, we will see a practical example of how to do this using machine learning.

While that particular example could be of direct use to some people, there are many other applications of AI for testing. In this section, we will analyze some other examples that will hopefully inspire you to create more tools that use AI for testing purposes (I would be proud and delighted to add those to future revisions of this book).

## Games

As we will see in *Chapter 11*, *How to Test a Time Machine (and Other Hard-to-Test Applications)*, games are one of the most difficult applications to test due to their indeterministic and dynamic nature. There are different points in which games could present a challenge and in which AI could help us with their testing.

For example, we could use AI to help us create exploratory tests, by using reinforcement learning, as we will see later in the *Sending your code to the Gym* section. This could help us create agents that will win the game for us or perform other particular actions within the games.

Selecting objects in games to interact with them is another particularly difficult task for automation to perform, so AI could help us on this front with the use of computer vision.

An example of useful AI for testing was done by Ubisoft. The company created a tool around 2018 named **Commit Assistant** *[4]* to try to catch mistakes in the game's code before the developers committed it. This was done by feeding the AI system with code where previous mistakes were made and predicting when the developers were at risk of committing similar ones. They claimed this tool would predict more than half of the bugs and save 20% of the developer's time.

## Data analysis

In the previous chapter, we talked about the importance of data science for testing. We learned how to normalize some of the data for the machine to understand it and analyze it automatically. However, automating how test data is analyzed could become difficult if it is written in natural language. For those cases, using NLP algorithms, we might be able to reach better data analysis, such as grouping failures to understand which pieces of the application are more problematic.

## Test case creation and/or automation

In a similar fashion as before, we could use AI to convert tests that are written in natural language, as is the case of using **behavior-driven development** (**BDD**) techniques to code (with NLP). Alternatively, we could do something similar to what GitHub's **Copilot** tool does – that is, generating test code requested by users' comments out of other pieces of similar code saved within the platform.

## Grouping tests

We could use different types of AI approaches to find commonalities between tests by their variables – for example, by their descriptions or steps definitions (using NLP to analyze them).

We could find useful variables and relationships that could help us improve our test case analyzer by adding the new variables to our inputs.

Another potential use of test grouping is to find repetitions of tests across the test pyramid by analyzing common steps, code, or other patterns across it.

## Automated visual testing

The new computer vision algorithms can help us automate manual visual checks by finding differences in expected versus actual images of the UI. The algorithms would take one or more partial or full screenshots of the state of an application or a website and validate them against a previous state. There are tools on the market that can help you decide which parts of the screen should be treated as static and/or dynamic for testing and with which accuracy to test each part, and even understand how the different parts should look across different resolutions and browsers. One such tool is Applitools.

We could find other tools that can perform visual-based testing as an addition to **document object model** (**DOM**) manipulation, being able to find specific objects visually within the application, rather than by the more familiar object's textual descriptors. An example of a tool that uses visual-based testing, among other techniques, is Netease's **airtest project** *[5]*.

## Smart DOM finder

As we mentioned previously, finding objects in the DOM could prove to be a difficult task. This includes objects without IDs, dynamically created, obscured by other elements, changing or 3D objects, and so on. Commonly, we end up using XPaths to find those objects which tend to change frequently, thus creating flakiness in our tests.

We could create a smart AI DOM finder, which would find the best-suited element from the available ones if the element is not found by the given textual descriptor. Alternatively, it could use a picture (also known as visual-based automation) to find that particular object visually.

## Testing by voice

Voice recognition tools are based on AI and could help us create, perform, or run tests without the need to click or type anywhere. This could improve the accessibility of the team (besides it being a cool project to do). An example of a tool that does coding by voice is serenade [6].

## Smart test balancer

Finding what test to run, and where and when to run it across the developing pipeline in an intelligent matter, could help us solve issues and reduce costs.

If you have the choice of running your test code on several servers, maybe across different time zones, you might want to find out the server and/or time that would reduce the number of users impacted or maximize the used resources, thus minimizing the cost while ensuring proper and timely results.

## Optical character recognition (OCR)

**Optical character recognition** (**OCR**) systems try to produce a text output from a visual input (such as an image) that contains written text.

OCR techniques could be helpful if your application deals with a lot of text that comes in the form of images. Depending on the language of the text, dedicated libraries could help you with this task (as an example, it is not the same to test English as Arabic or Chinese since characters, the relationships between them, and even the direction of the text might vary). However, if what you are checking does not have to do so much with the content's meaning as with the shape of it overall (as in, where the text is placed rather than its meaning), it might be more convenient to use some of the already mentioned libraries for visual testing or visual-based testing (screenshot testing) or a combination between them and OCR libraries.

In the next section, we will deep dive into our test case analysis tool again, this time using machine learning. Hopefully, this will inspire you to create some other AI testing tools that we could mention in a future update of this book!

## Test case analysis with machine learning

Regarding the way that we built our test case analyzer in the previous chapter, it is argued not to be part of AI as an expert needs to give the computer specific instructions or rules for the system to work, which makes it not as "intelligent" as if it was to learn without explicit instructions. For instance, if we take the second definition of AI that was provided in this chapter, we can say this is not strictly AI as the computer is not implicitly learning the rules.

Many years ago, using machine learning for AI seemed unreachable with the hardware at the time, so most of the research and tools were being done/created using some sort of rule based system. However, with the new systems being able to compute everything much faster and the creation of better algorithms that this made possible, research within AI is being moved toward machine learning and other more automated learnings instead.

Before moving on to further applications of AI that could be useful for testing, let us see how we would complete the example in the previous chapter by using machine learning instead and make a comparison of the two cases. In this case, for contrast, we will use a Python library called `sklearn.neural_network`[7].

The following subsections explain the steps involved in this process.

### Getting a training set and a test set

The next step is to separate our data into training and test data. The training data is used by the algorithm to figure out the best solutions (this is called the training phase). The test data is used to evaluate whether the algorithm works correctly (this is also called the inference phase).

Instead of using some of the final inputs that we want to get the outputs for, we prefer taking inputs from part of the actual known data so that we know what outputs should be produced as a result; that is why we split the known data into two sections (one to train the system and the other one to validate the training). Then, we can validate the outputs obtained by the trained system with the real ones, so that we understand if the learning was done appropriately. Lucky for us, the **sklearn** [7] library has a function for splitting data for this purpose (`train_test_split`). We can give it a random seed (`random_state`) to make sure the splits are reproducible every time we run the code and that we can understand how the changes in the algorithm's parameters affect the overall results. We can also indicate the percentage of data that we want to use for testing with `test_size`. Feel free to play around with these values and print or save the training and test sets to a document to understand how they work better.

Let us see how this is done:

---

**neuralNetworkAnalyzer.py**

```python
from sklearn.model_selection import train_test_split
#[... import pandas and create dataFrame as in
#dataNormalizationSKLearn.py ]
# saving the expected data as a target - we prepare it for
# later processing
target_col = ['Expected']
# Adding the variables into their own grid for later use -
# for later processing
variables = list(set(list(dataFrame.columns))-set(target_col))
# Creating training and test datasets
X = dataFrame[variables].values
# using ravel to transform to array
y = dataFrame[target_col].values.ravel()
X_train, X_test, y_train, y_test = train_test_split(X, y,
stratify=y, test_size=0.05, random_state=42)
```

### Training the system

Our system is now ready for the actual machine learning training. We will use a method called `MLPClassifier`, but please note there are many libraries and methods you could use for this. **MLP** stands for **multilayer perceptron**, which means that it will use multiple layers to train the system, which we can configure with the `hidden_layer_sizes` parameter. The `activation`, `solver`, and `max_iter` parameters also affect the overall settings of our algorithm. Please check the documentation of the sklearn library for more information on these.

Let us learn how to train the system:

---

**neuralNetworkAnalyzer.py**

```python
# Training the system
from sklearn.neural_network import MLPClassifier
classifier = MLPClassifier(hidden_layer_sizes=(400, 200,
100), activation='relu', solver='adam', max_iter=300, random_
state=42)
```

```
classifier.fit(X_train, y_train)
predict_train = classifier.predict(X_train)
predict_test = classifier.predict(X_test)
# here we could add another classifier.predict with the actual
data we want the algorithm to perform
```

### Analyzing data results

As mentioned previously, the algorithm we've used allows for certain parameterization. No algorithm is perfect and, commonly, we need to find the right number of layers and weights and fit all the variables to get the best results for our system. To do so, the first step is to analyze the data and determine whether the results are performing well. We could do this as in the previous chapter, outputting the results versus the expected results and comparing the percentages of good matches. However, with sklearn, we have two methods that can help us visualize and get a report of the precision of our system: confusion_matrix and classification_report.

In the following piece of code, we can see how well the system performed for the training model and the test model:

**neuralNetworkAnalyzer.py**

```
# Analyzing the data on the train model
from sklearn.metrics import confusion_matrix, classification_
report
print(confusion_matrix(y_train, predict_train))
print(classification_report(y_train, predict_train))
# Analyzing the data on the test model
print(confusion_matrix(y_test, predict_test))
print(classification_report(y_test, predict_test))
```

The output is as follows:

```
[[8939 4826]
 [5282 8503]]
                 precision    recall  f1-score    support

          0.0        0.63      0.65      0.64      13765
          1.0        0.64      0.62      0.63      13785

     accuracy                            0.63      27550
    macro avg        0.63      0.63      0.63      27550
 weighted avg        0.63      0.63      0.63      27550

[[377 347]
 [378 348]]
                 precision    recall  f1-score    support

          0.0        0.50      0.52      0.51        724
          1.0        0.50      0.48      0.49        726

     accuracy                            0.50       1450
    macro avg        0.50      0.50      0.50       1450
 weighted avg        0.50      0.50      0.50       1450
```

Figure 8.1: Output of the confusion matrix and classification report
for training first, then similar output for test

Note that for this example, the data that we used was auto-generated randomly, so it is normal that the results are not the best. We hope to get above 60% (0.6), hopefully toward the 80% (0.8) success in the f1-score. However, if we happen to find something like this, we should look into improving these values by changing the number of layers and the other parameters of the algorithm.

The two numbers in the last part of the confusion matrix mean that *we caught 348 actual issues and missed 378*. This also seems very low for this case. Again, this is normal since the test cases and issues were randomly generated and I didn't tweak the algorithm to get better results, but in a normal system, we will see some commonalities that allow us to *create rules* and avoid us having to execute all of the tests to get the same results.

## Considerations and comparing the two systems

Comparing the two systems is not a straightforward task. First of all, one of the benefits of a rule-based system is that we do not need as much previous data as with machine learning, as there is no training phase. This means that an honest comparison would be of detriment for one or the other system, depending on how much previous data we before hand to train the system.

To run a comparison of the two systems, we should first find a way of understanding how well they performed. For this, we could get a snapshot of the system and check how well the system is capable of predicting the output for that snapshot – for example, by taking out the last result and checking it matches the predicted value. This is not the best measuring form, and it would be interesting to keep measuring the system as it grows or keep training it every so often to ensure its accuracy. In our example, we gave random values to the expected value for us to exemplify this step.

In both systems, we (humans) decided on the variables that we thought were important (supervised learning) since we did not have a lot of variables to begin with. However, we may want to run some unsupervised learning to figure them out, to avoid cases in which some of the variables could be measuring the same things or relate to each other in a way that might not appear obvious to a human.

For example, a human might mark a new and risky feature for automating and testing, but our AI system might find this is not needed. When analyzing the result, we might learn that a possible reason is that our developers put more effort into the code of the new features, which is something we could have not realized of on our own. (*Note: This is hypothetical, not necessarily the case – it is a good idea to cover risky features, even if an AI tool tells you not to.*)

As in the previous chapter, with the rule system, we might influence the results of our system highly by the rules we explicitly specify. These rules are also difficult to define, especially when they have a fuzzy nature. Finally, another aspect of defining rules is that they are static; we need to manually add new rules if the situation changes and re-evaluate the system. However, some other AI systems get to change and adapt automatically to their circumstances.

Long training sets could take longer to train than manually setting rules. We may also overtrain the system (which means that we skip the best value after too many iterations or because we iterate using too big steps per iteration). If the data changes a lot, the machine learning system could lose effectiveness too, although the same will probably happen when using rules.

This can be summarized in the following table:

| Rule-Based System | Machine Learning |
| --- | --- |
| Does not need much data to get started | Works better with bigger amounts of data |
| Need to manually create rules that might be influenced by us and fuzzy definitions | No need to set rules |
| We specify the variables | We could make an unsupervised learning system to get the variables |
| Time and experience needed to manually identify and set rules | The time for training could be long and we could run into possible overtraining or overfitting |

Table 8.1: Summary comparison between the systems

On the other hand, both systems can help us highlight issues with the test case plan. For example, do we really want to perform a test if it's always a failure?

The number of steps could be converted into a total expected time by taking an average of $X$ seconds to run or to automate per step (whichever time you consider would suit your system best). This could give us a way of retrieving the maximum tests that could fit that time.

In the preceding examples, we are lacking information on retrieving the test cases that we marked as "to execute or automate" from the training set (this would take an iteration over those test cases finding the *expected* value to be *true*).

In this section, we reviewed some uses of AI in testing. In the next section, we will identify different types of AI applications and what we should do on the test side for them.

# Testing AI apps

Now that you have a better understanding of the different types of learning and AI approaches, you should be able to guess that there are just as many different types of AI applications. In the previous section, we saw some of those types when we reviewed using AI for creating apps for testing.

## Types of AI apps

Let us group them into the different types of learning so that we can think of ways of testing them.

### Supervised learning apps

As we mentioned previously, these applications get the inputs and outputs to produce a result. Classifiers, such as the test case analyzer that we reviewed in the previous section, are examples of these types of apps.

In these apps, the AI test itself is done as part of the AI automation phase. Therefore, there is nothing to do in this aspect from *a quality team's* perspective. The app should be tested as any other application, checking for performance, accessibility, usability, and every other aspect of the app itself that is not the AI bit or the results obtained with it. However, if we want to double-check the development of the AI, we could also have some small checks to make sure that the AI still works as expected and that it has been trained appropriately because sometimes, the test data might be biased toward certain values. We can even review that data ourselves to verify that the quality of the test set is good enough, that the produced results look good or meet some criteria.

Following the example of the test case analyzer, it might be difficult to understand what process should be used to test this app as this is mostly *back-end*. So, let us look at a different example: a **number plate recognition** (**NPR**) system. NPR systems could be considered a specific type of OCR as they recognize the characters of a number plate from a visual input. OCR systems such as this one generally land under the definition of supervised learning.

The back-end of the system is the AI behind recognizing the number plate itself. But that is not the only part of this app. The app could be designed for different purposes, and each of them would have a different *front-end* that we should make sure to test. Let us imagine we use NPR for a car inspection. There would be a way for the user to upload the number plate or full car picture into the system (it could be automatically set up within the hardware of the center). Then, the system would provide some output, maybe a form with details of the inspection to fill up.

All those UI fields should also be tested and ensure that the NPR field gets filled up after taking a picture with the correct number of the car's number plate. We should also test for front-end expectations when the NPR could not identify the image.

We should perform any other type of testing that applies as well (performance, accessibility, security, and so on).

Lastly, if we want to make sure that the system is properly trained, we could try a bit of chaos testing: different typographies, tilted plates, different distances of the taken picture, and so on. However, this is expected to have been already tested by the people who built the AI system in the first place, so, hopefully, it should not be too much of a concern for us.

### Unsupervised learning apps

These apps are useful when we do not know what exactly we are looking for in data (categorization or clustering), or we do not require training data to be labeled. An example could be personalized recommendations based on your and other people's histories.

How could you possibly tell that the recommendations that the AI application is giving you are correct without feeding the algorithm with test data? We won't be diving into too much detail about this, but this should be the responsibility of the person who writes the AI code. There are a series of mathematical scores and error calculators that would ensure that the categories produced are the best ones achievable.

As with the previous example, we could find the AI logic to be embedded as part of a bigger system that we will have to test as we do for any other app, trusting the AI logic to be tested as part of its development. However, we can make some checks to verify that the produced results look good or meet some criteria.

Besides this, the impact on sales or usage after the application is in place can also give the business an idea of its success, although this success may as well happen by chance due to some other factor not related to the AI system. Therefore, it is important to keep track of all the variables involved.

## Considerations about AI apps

There are some important aspects of AI apps that we should be careful about.

## Too much accuracy

Even if this might sound counterintuitive, too much accuracy could be a bad thing. For instance, we might get a system that can tell what the user wants before they even know they want it. An example of this could be a shopping mall predicting a pregnancy. If privacy is not guaranteed for those systems, the user might not be happy about these predictions being shared even to their household.

## Continual learning

While most applications contain static data and can use a unique phase of learning, some applications require constant refreshing of their learning to guarantee accurate results over changing data. These applications tend to be trickier to test and we need to ensure that the passed data is correct; otherwise, it could be badly trained. If users provide the scores for success and failure, the system could be opened to users that train it badly on purpose or because of misunderstandings about the scoring.

Another problem with continual learning is that we do not have control over what the apps are learning from, which leads to the concern of machines possibly gaining consciousness if they have the power to and can collect any data without restrictions. This can lead to arguments about the level of influence that the test team should have over the ethics and designs of an application and if they should be the ones to stop it if they feel the development has gone too far.

Finally, if you were to test an app that is continuously learning, you could try using a copy of the state of each version to test in an isolated environment so that the test data, for whichever type of test you are performing, does not influence the overall experience for the users. Also, we should be careful about the different types of testing we perform and whether they might add a learning iteration to the AI part.

## Other issues

Other issues might arise from AI apps that we should be aware of and make sure all the team is too. We have already discussed a bit about the importance of having good data for training to avoid biased data. Furthermore, we should make sure that the app works for every person that needs to use it (for example, different names or faces should still work, different types of login should work if users can't use face or fingerprint ones, and so on).

Besides this, we should make sure that the physical security of the users is not compromised if the apps deal with hardware (such as in the case of autopilot for a vehicle).

In this section, we looked at different examples of AI apps and considerations on how to test them. In the next section, we will learn how to use reinforcement learning to create some AI testing tools.

# Sending your code to the Gym

**Gym** *[8]* is Python's toolkit for reinforcement learning algorithms. This tool is commonly used for solving games, which could help us if we were to create test automation for games. However, it is not the only reinforcement learning tool that exists and games are not the only potential uses for it. In the previous chapter, we created an exploratory web crawler. We could create another version using reinforcement learning so that instead of telling the system where to crawl next, the system will crawl to the areas where crashes are most probable.

Creating a crawler using reinforcement learning might not be the best solution for it, but this is only an example to help us explain this concept in a friendly fashion to qualified experts and to inspire you to think out of the box concerning testing. Furthermore, keep in mind that there are some tools out there that already have crawlers built in, and you might be better off exploring purchasing such tools rather than creating your own design from scratch.

Each programming language has reinforcement learning libraries. In this section, we will focus on Python and Gym; other alternatives are for you to explore.

First, let us learn more about the Gym library with a basic example so we can estimate how we could use it for the web crawler so that you are able to play around with this, or other projects, or other similar libraries on your own.

The following three main elements of Gym are required for an intelligent agent to learn by reinforcement:

- **Reward**: This is a float number that represents how well the previous action was performed by the intelligent agent. The goal of the agent is to maximize it.

- **Observation**: This is an object that represents what the agent knows about the environment. Later in this chapter, we will define a system with different possible states and use this value as our new state. This could also be pixel data or the position or rotation of a player.

- **Done**: This is a Boolean value that becomes true if the environment should be reset – for example, if the goal is achieved or missed.

Gym has a fourth element, which is a dictionary of information for debugging purposes.

You can find out more about this library at `https://gym.openai.com/`.

You can install it with the following command:

```
pip install gym
```

Note that you might need to install `gym[toy_text]` in the same manner.

In our example, we will use the following libraries:

## gymExample.py (imports)

```
import numpy as np
import gym
```

Here, NumPy is a library for scientific computation in Python and Gym is our reinforcement learning library.

The first step in reinforcement learning is to define the environment that the agent is going to be in. For games, this means defining the board and location of each of the components of the game. In the case of the crawler, we can define the environment by URL and actions to navigate to the different links.

For a basic example, we are going to use an environment called `frozen lake`, which can be found at *[9]* in the *Further reading* section.

This environment creates a random text map of letters – F, H, S, and G:

- S represents the start of the map

- G represents the goal or end of the map

- F represents a frozen portion of the lake

- H represents a hole in the lake

The goal of the intelligent agent is to reach the goal from the start without falling into any holes. In this environment, there are four possible actions: going left, right, up, or down. Let us see this visually:

## FrozenLake-v0 (visual representation)

```
SFFH
FHFH
FFHH
HFFG
```

This environment is loaded with the following instruction:

## gymExample.py (environment loading section)

```
environment = gym.make('FrozenLake-v0')
environment.reset()
environment.render()
```

How can this relate to or represent our crawler? We could have a similar environment for our application, in which every step is a potential way to navigate from the current view to the next. In every step, we store the number of defects (bugs) that have been previously found on that view. Our agent would have to find the route that would collect the highest number of defects.

One complication with the crawler's environment's representation concerning the frozen lake is that each of the navigations could lead to several other navigations. The actions are actual view redirections rather than the four stated actions (going left, right, up, or down). The start point can easily be found – that is, the index, home page, or start view. However, the ending point could be complicated as we might not have a final URL, which could happen with a website.

To sort out this problem for our example, we can set a maximum number of redirections to stop the system from going on indefinitely, rather than a final destination (although that could be a different design):

## A potential example of a crawler environment for a website

```
URL             Nbugs found         Links from page
https://www.packtpub.com/    7               browse_all,
free_learning, user_sign_in, cart, subscribe, ...
browse_all               19              book1,
book2,....
.....
```

Another problem with this environment is that, in the case of the crawler, technically, we do not know the URLs related to a view until we go in there. The same goes for the number of known issues. There are a number of different approaches we could take to design this system:

- The data to define the environment could be gathered and modified by hand if the website design is not too complicated.

- We can have a crawler that goes through all the URLs once in order to retrieve the details of the application and save them into our environment. Then, we can use the crawler with reinforcement learning as a BVT of sorts, just going through the more error-prone sections and retrieving faster results. Regarding the number of defects, these could be retrieved from the previous crawler or pulled from our defect tracking system.

- We could have a dynamic environment that gets created as the crawler works and adds defects as it finds them. This solution is a bit more complicated on the technical front than the previous one as it might require a different algorithm than the one that we will show in this chapter. It will also need to call a function to find out issues on each of the views as it crawls the system (rather than using the ones provided in the table).

Feel free to experiment and create environments for this or any other solution that you might need.

To solve our problem, we will be using an algorithm called "Q-learning." Feel free to search for other algorithms or find out more about this one.

In our algorithm, when the agent goes around the system, it will analyze every possible result after applying each action to the current state, and it will build a table with the best values of each of the iterations so that we can keep track of them. This is generally called a **q-table**, but we will call it a **bug-table** as it fits our logic better.

First, the agent will have a 0 value in all the cells, and in each state, the value would be updated with a positive or negative reward, depending on what is found in each cell.

In the case of the frozen lake, that would involve adding 1 for F and subtracting 1 for H. In the case of the crawler, we would have to add the number of defects known for that URL, which would be specified on the bug-table.

This table can be initialized in the following manner:

## gymExample.py (Table loading section)

```
observation_size = environment.observation_space.n
action_size = environment.action_space.n
bug_table = np.zeros((observation_size, action_size))
print(bug_table)
```

Here, `environment.observation.n` and `environment.action_space.n` indicate the number of states (4: S, F, H, G) and actions (4: `down`, `up`, `left`, `right`) in the loaded environment, respectively.

You can check what values this table contains at any point with the `print(bug_table)` instruction.

Now, it is time to solve the problem, for which we would need a specific algorithm. While some solutions and algorithms do not require our `bug_table`, it would be nice to update such a table to track the state of our system. We recommend trying different ones for your system and finding which will make more sense for it.

In our algorithm, we will iterate a maximum number of times. Each iteration is like the AI trying to reach the goal from a different path or angle. Therefore, this is setting up the maximum number of times the AI can try to go through the game (or the website) to find the best solution.

General algorithms to solve exercises such as the frozen lake have a goal to find. However, in our case, our goal does not exist, so we must depend on the number of iterations to stop the algorithm. Therefore, we need to make sure we have a rough idea of how long the algorithm would need to achieve good results. Fitting algorithms in AI sometimes comes down to trial and error. We could potentially set a final link to use as a goal, depending on what our link tree looks like, and leave the algorithm to reach it.

If there are no links in the view or page we are visiting, we must go back to the initial one.

Every time we want to perform an action, we must use the `observation, reward, done, debug_info = environment.step(action)` instruction to perform it, and then `environment.render()` to update our environment with the performed action.

A simple random solution could be implemented by iteratively retrieving the next actions until we find the goal or we reach a total number of iterations, `TOTAL_ITERATIONS`, so that we do not run into an infinite loop. This solution looks as follows:

## gymSimpleSolution.py (Algorithm section)

```
for i in range(TOTAL_ITERATIONS):
    action = environment.action_space.sample()
    observation, reward, done, degub_info =\
        environment.step(action)
    environment.render()
    if done:
        break
```

The loop will finish when `TOTAL_ITERATIONS` is reached.

This sort of solution might be sufficient if we are just finding a goal. However, since we want to maximize the number of defects found, we must take the best action from the possible ones and consider our table of defects.

For the lake example, the action to take must be the one that would guarantee us to find the final state in the shortest amount of time. For the crawler example, the action to take to exploit the system must be the one that will maximize the number of defects that we could potentially find in the maximum number of iterations.

In our algorithm, we will have a probability of the action being explorative (random) or exploitative (maximizing results). We are following an algorithm that's a bit complicated but has proven to work well. This algorithm has a way of updating this probability, but we could always have the same probability instead, for simplification.

Then, we should update our bug-table after we perform every action. The algorithm follows an equation that has several constants to perform this update and predict if the next step will be better than the current one.

Finally, in this case, we must consider that, for our website, all of the pages will have a natural redirection to the first page if no other URL is linked to them.

Before jumping into the algorithm, let us review what is supposed to happen in the pseudocode:

## gymExample pseudocode

```
for i in range(TOTAL_ITERATIONS):
    current_state = env.reset()
    done = False
    current_total_reward = 0
for i in range(TOTAL_REDIRECTIONS):
    if !exploration:
        # Exploit the system - take the best action
        action = np.argmax(bug_table[current_state, :])
      else:
        # Explore the system - take a random action
        action = env.action_space.sample()
    observation, reward, done, degub_info =\
        env.step(action)
    new_estimated_value =\
      calculate_estimation_reward_next_step_vs_current()
    bug_table[current_state, action] =\
        new_estimated_value
    total_rewards.append(current_total_reward)
    current_state = observation
    if done:
        break
total_rewards.append(current_total_reward)
environment.render()
```

We have highlighted what is missing from this pseudocode to be actual code. First, to check whether we should explore or exploit, we must use a probability. This probability starts with a fixed number (which is one of the values that we can tweak to improve the algorithm). Then, it needs to be re-calculated in every loop so that the probability of exploitation increases as we loop through the iterations.

The second part involves updating the bug table. In this case, we must also use AI to try to predict whether it would be better to take any of the steps we are evaluating to proceed, or whether it would be better to stay where we are. We do this calculation by using some sort of algorithm that would make such a prediction. In our case, that's part of the q-algorithm. As before, this one will have some variables that we can tweak to get better results.

To understand the "tweaking" of "fine tuning", remember what we covered when we covered the test case analyzer. AI algorithms tend to contain some variables that can be adjusted to find the best results for each case and algorithm. In AI terms, this is called "fitting" the algorithm.

In the following example, we can see what this algorithm looks like. Check the commented lines to understand what each section does. Note that this code will be using several constants that need to be fit to improve the algorithm. First, let us look at an example:

## gymExample.py (constant definition)

```
EXPLORATION_PROBABILITY = 0.5
EXPLORATION_DECREASING_DECAY = 0.001
MINIMUM_EXPLORATION_PROBABILITY = 0.01
DISCOUNTED_FACTOR = 0.1 # gamma
LEARNING_RATE = 0.99
```

Now, let us see how the algorithm will work:

## gymExample.py (algorithm section)

```
total_rewards = []
for i in range(TOTAL_ITERATIONS):
    current_state = env.reset()
    done = False
    current_total_reward = 0
for i in range(TOTAL_REDIRECTIONS):
# We iterate through the number of redirections allowed
# from this step we can set this as a constant or as part
# of the table
# Next step is to decide if we want to exploit or explore
# the system
    if (np.random.uniform(0,1) >\
        exploration_probability):
        # Exploit the system - take the best action
        action = np.argmax(bug_table[current_state, :])
    else:
        # Explore the system - take a random action
        action = environment.action_space.sample()
    # update the system
```

```
    observation, reward, done, degub_info =\
        env.step(action)
    # calculate the estimation of next step - this is
    # Done from the formula of the q-algorithm
    discounted_estimated_next_step =\
        learning_rate * (reward + discounted_factor*
                        np.max(bug_table[observation,
                        :]))
    # update the bug table with the new values (based
    # on the q-algorithm formula as well
    bug_table[current_state, action] =\
        (1-learning_rate) *
         bug_table[current_state, action] +
         discounted_estimated_next_step
    # update the total reward
    current_total_reward = current_total_reward +\
                        reward

    # go to next step
    current_state = observation
    # stop if done
    if done:
        break
# add the new reward
total_rewards.append(current_total_reward)
# We should now fix the probability of exploration, based
# on the algorithm as well, but we could have always the
# same one
exploration_probability = np.max(min_exploration_prob, np.exp(-
exploration_decreasing_decay*2))
print(bug_table)
# Finally, we render the environment that we got
environment.render()
```

This algorithm should find a solution for the frozen lake problem. If we have an environment for our website, we could try to use it to maximize the number of defects. Ideally, we should add the capability for it to find actual defects as it goes, by calling some functions before the table arrangement. We will not include this part in this chapter to avoid overcomplicating the code.

Also, note that this would be much less efficient than a normal crawler as it will need to iterate over the website, thus it also introduces false traffic, which could affect the users if done in production. However, once we have the solution, we could have yet another crawler that goes through it in a BVT or sanity-test fashion. Whether that would be useful or not depends on your system's needs. We just wanted to give you an example to help you understand reinforcement training in a manner that is familiar to you.

In this section, we learned how to use reinforcement learning to assign weights of testing importance to the different parts of an application. In the next section, we will cover other projects that you could use to learn more about AI and testing, and some tools that are currently available on the market that could be of use in your system.

# Other cool AI projects and tools

AI tools and projects are growing exponentially, so, by the time you read this chapter, there will probably be many more than the ones discussed here. However, we will look at some of them here to get you started. In this section, we will explore other tools for AI, then discuss ways of learning more about this in case you want to get more general knowledge to create your own tools.

## Other tools for AI

While working on previous projects could help you gain more knowledge of AI and how it can be related to testing, there are already many tools on the market that you could use directly to aid you with your AI-related issues. Instead of reinventing the wheel, try looking around and finding the one that could be adjusted to meet your specific needs. We mentioned some of them in the *AI for testing* section of this chapter.

We can recommend Python as the language to use if you are starting with AI tooling due to its ease of use and small learning curve and the number of existing libraries for it that support AI. However, there are many other languages (such as most high-level programming languages) that support AI programming.

If you are working with Python, make sure you check out **TensorFlow** *[10]* or **PyTorch** *[11]* for your AI needs. You can also try **Jax** *[12]* or **sklearn**, as we have shown in the coding bits throughout this chapter. **NumPy** *[13]* and **pandas** are libraries of help for numeric manipulation and matrixes. **Pillow** *[14]* is a good one to know for handling images. **OpenCV** *[15]* also has a library for Python, which will help you with computer vision projects. Remember to try **Gym** for any reinforcement learning, especially related to games or problems with a defined environment, along with starting and goal states.

If you are unsure about which library to learn from, feel free to learn more about each of them by exploring their documentation and trying a small project using each one. There are many more than the ones listed for you to explore and try out.

If you are practicing or developing any particular side of the AI spectrum, make sure you also read some theory about that aspect of AI and check out existing algorithms. You won't likely need to create an algorithm of your own, but you might need to use some to find the best solution, as we did in the previous section's exercise, rather than trying around randomly and blindly..

We have already mentioned some tools that use AI and assist with testing. Note that I do not want to advertise any particular tool; instead, we will just talk about those we have tried out at some point, but there might be equivalent tools on the market that will help you as much. Make sure you do an extensive and updated review online before deciding on any of them.

## How to learn more about AI

As we mentioned at the beginning of this chapter, AI is something that experts dedicate their entire careers to. Not to downplay their job and knowledge, but there is still much you could do to learn about AI and many courses available in each of the areas. This knowledge would help you use, understand, and even create (by yourself or with the help of AI experts) specific tools that are related to testing, helping you get some expertise on this side of testing.

If you want to gain more knowledge in the AI area, you should have a clear understanding of what you would like to learn about. If you are still not sure about this after reading this chapter, you should start with some introductory courses, then advance toward the specific materials that you would need for whatever you are interested in.

Once you know what area you want to gain more expertise in, you can find resources such as books and courses. There are also multiple projects and competitions that you can join to try to get more hands-on with AI (as was the case with Minecraft AI programming with Project Malmo, or Go-solving competitions).

Keep in mind that this, as happens with many areas in computer science, is a very dynamic subject, so keeping your knowledge up to date is advisable.

# Summary

In this chapter, we reviewed AI, a concept that is getting more and more popular across all computing areas as more powerful machines and new algorithms that use that power get developed.

We started this chapter by exploring the foundations of AI and why it's important to know about this. Then, we reviewed some general knowledge about technical concepts related to AI.

We also saw some AI tools that could be of use for testing and learning about types of AI apps and the specifics of each type when it comes to testing them.

We completed the analyzer that we reviewed in the previous chapter with machine learning and looked at the difference between the two systems and when you should opt for one or the other.

We also looked at some examples related to reinforcement learning, which could be useful for finding solutions in games and other applications.

Finally, we talked briefly about tools for AI and other projects that could be of interest.

In the next chapter, we will review another hot topic from a few years back that is now everywhere to the point of becoming mainstream: the cloud.

## Further reading

To learn more about the topics that were covered in this chapter, take a look at the following resources:

- [1] More about GitHub's Copilot tool can be found here: `https://github.com/features/copilot`.

- [2] More about ChatGPT can be found here: `https://openai.com/blog/chatgpt`

- [3] More information about Python's pandas library can be found here: `https://pandas.pydata.org/`. There are also some books on our website regarding this subject: `https://subscription.packtpub.com/search?query=pandas`.

- [4] See the presentation of Ubisoft's Commit Assistant: `https://montreal.ubisoft.com/en/ubisoft-la-forge-presents-the-commit-assistant/`.

- [5] Netease's airtest project: `http://airtest.netease.com/`.

- [6] More about serenade at: `https://serenade.ai/`

- [7] More about the scikit-learn library: `https://scikit-learn.org/stable/index.html`.

- [8] The main GitHub page for the Gym toolkit can be found here: `https://github.com/openai/gym`. The documentation site is `https://www.gymlibrary.dev/`.

- [9] The frozen lake example for this chapter has been deprecated, but this and the new version can be found here: `https://github.com/openai/gym/blob/master/gym/envs/toy_text/frozen_lake.py`.

- [10] For more information on TensorFlow, visit `https://www.tensorflow.org/`. You can find some books about it on our website: `https://subscription.packtpub.com/search?query=tensorflow` (you can also find some interesting courses on different platforms online).

- [11] For more information on PyTorch, visit `https://pytorch.org/`. You can find some books about it on our website: `https://subscription.packtpub.com/search?query=pytorch` (there are also some interesting courses available online on your preferred coursing websites).

- [12] Documentation about Jax can be found at `https://jax.readthedocs.io/en/latest/notebooks/quickstart.html` and `https://github.com/google/jax`.

- [13] You can find out more about NumPy on their website at `https://numpy.org/`. You can also find some books on Packt's website: `https://subscription.packtpub.com/search?query=numpy`.

- [14] You can find out more about pillow at `https://python-pillow.org/`.

- [15] You can find out more about computer vision and OpenCV at `https://opencv.org/`. You can also find some books on our website: `https://subscription.packtpub.com/search?query=opencv`.

- [16] If you want to find out more information about the other concepts presented in this chapter, check out the following links:

  - Machine learning: `https://subscription.packtpub.com/search?query=machine%2520learning`

  - Artificial intelligence: `https://subscription.packtpub.com/search?query=artificial%2520intelligence`

  - Deep learning: `https://subscription.packtpub.com/search?query=deep%2520learning`

  - Computer vision: `https://subscription.packtpub.com/search?query=computer%2520vision`

Don't forget that you can also find other courses online on your favorite platforms.

# Having Your Head up in the Clouds

The cloud has been a hot topic for many years, to the point of becoming mainstream (everybody is using it in one way or another). In this chapter, we will see some concepts about the cloud, how it can help increase the quality and development speed of your apps, and which things you have to be careful about while using it.

Some people might say that *having your head up in the clouds* is a bad thing, but in this case, in regard to testing, I believe it is something you should definitely aim for.

In this chapter, we will cover the following concepts:

- What exactly is the cloud, and how can it help you?
- Creating a benchmark to measure testing performance
- Testing appropriately in the cloud
- Dangers of the cloud
- Testing applications in the cloud
- Thinking out of the box about the cloud

## Technical requirements

Some degree of programming skills is recommended to get the best of the examples provided in the chapter.

In this chapter, we will use a variety of programming languages (mainly Java), with TestNG as the unit testing framework for Java.

We recommend reviewing and working with different languages as a self-growth exercise, and we provide some examples in our GitHub repository, so it should be easier for you to play with them: `https://github.com/PacktPublishing/How-to-Test-a-Time-Machine/tree/main/Chapter09`.

While this chapter is written with QA/SDET roles in mind, developers may also find this one interesting and useful, especially when selecting cloud testing providers or testing serverless applications.

## What exactly is the cloud, and how can it help you?

Before the 2010s, if you'd heard the word *cloud*, you would have imagined some sort of cluster of soft-looking water in a gas state floating in the sky. However, nowadays, when you hear *cloud*, you might imagine some application that allows you to store data or use some device without the need for local ones. A cluster of powerful computers or other devices stored in multiple secure undisclosed locations might also come to your imagination, as this is how the services are implemented in reality.

Whichever the definition, for most of the 2010s, it was *the hot topic*, and everybody seemed to want things to happen in the cloud. And that's how mostly everything was reborn, by adding "...in the cloud" after it. Your data? In the cloud. Your applications? In the cloud. Your testing? In the cloud!

Even though enthusiasm about the topic sounds slightly colder now, since many companies are providing their own clouds and renaming them, the cloud definitely seems to be here to stay. In fact, it is far from frozen as it keeps slowly growing, and new capabilities and uses for it keep being invented.

As we will discuss in this chapter, the cloud can help you in many ways and areas, including during testing itself. One big benefit of using cloud services for testing, besides being able to reach more devices and locations than your budget might allow you to otherwise, is test parallelization.

In *Chapter 4, The Secret Passages of the Test Pyramid – The Top of the Pyramid*, we reviewed the need for test parallelization (including some associated dangers) and remote execution. In fact, testing in the cloud is not much different than testing on a remote server, taking some specific capabilities or syntax from your test provider. And as we will see in this chapter, parallelizing tests in the cloud could save you time and money.

Companies that offer cloud testing solutions have a group of devices across different locations that are generally referred to as a "farm" of devices. This allows the client companies to reduce their cost of purchasing and maintaining such devices so that they can focus on their products while providing good coverage for their customers and enabling more of them to use their services.

If you are interested in running tests in the cloud, several companies offer such a service. Without any particular order, these include **Amazon Web Services** (**AWS**) Device Farm (see *[1]* in the *Further reading* section), **BrowserStack** *[2]*, **HeadSpin** *[3]*, **Sauce Labs** *[4]*, and so on.

I happen to have been involved in some way or another with all of the aforementioned services; however, I feel a comparison of them from my side would be biased. In fact, if you find one such comparison, I would still not recommend you believe it right away.

The best way to proceed is, therefore, to make sure the provider is the right one for you. Here are some ways you could achieve that:

1.  Inform yourself of the features that they provide and check if they align with your needs:

    A.  Are the devices simulated or real? Sometimes simulation might not be sufficient for testing your application.

    B.  How many people are the devices shared with? Sharing resources might result in longer waiting time for you to run your tests.

    C.  What are the cleanup systems in place between executions? Otherwise, your data and application might be shared, so you need to provide the cleanup yourself (it could be a good idea to do so anyways, just in case, but your tests will take a little longer).

    D.  Which devices are supported?

    E.  What are the programming languages supported?

    F.  Are the tests uploaded to the devices or run remotely? Uploaded tests mean that your test code navigates through the internet and there could be a security concern.

    G.  Speed:

        • How long does it take to run a single test on average?

        • What's the longest test that it can run before a timeout occurs?

        • How long does it take to obtain a device/machine to run the test?

        • How long does it take to upload the necessary data?

        • How long does it take to perform the cleanup?

    H.  Which solutions/what logic do they have in place to select a machine?

    I.  What sorts of logs do they provide? Do they provide video recordings of the test execution?

    J.  Which time zones are covered?

    K.  What are the limitations? Per machine use or per of use?

    L.  What are the details of disaster recovery? How long does it take to sort out issues?

    M.  What sort of support is provided?

    N.  How secure is the system?

    O.  What potential integrations are there? (With **continuous integration/continuous deployment (CI/CD)** systems, for example)

2.  Ask for pricing or deals and get trials with the ones that match your requirements the better.

3.  Run benchmarks to validate their execution speeds for your particular system.

4.  Select the one with better benchmark results, pricing, and requirement matching.

In the next section, we will cover how to create a simple benchmark to measure results, but you should always tailor it to your needs.

# Creating a benchmark to measure testing performance

In *step 3* of the previous list, we mentioned running a benchmark, which is a piece of code or an algorithm that measures something. In this case, we can measure the time that it takes to run each test on each platform and compare them. The performance measured might be a deciding factor for the platform to use but might not be the only one, as there are other important aspects in the previous list, in *steps 1* and *2*.

In our case, we will measure the time it takes to run each test and the time it takes to run the entire class. Some cloud providers add extra time to their system setup and teardown or cleanup, and you should be aware of these values too. Sometimes, there is also some extra time taken if you need to send the test for it to be uploaded to their system (as opposed to running it remotely).

Let us see an example of how to make a simple benchmark with two tests using Java and TestNG (don't mind the code repetition in the actual tests in the example, but pay attention to the logging):

**benchmark.java**

```
package testingTimeMachines.chapter9.Benchmark;
import org.testng.annotations.BeforeClass;
import org.testng.annotations.AfterClass;
import org.testng.annotations.Test;

public class Benchmark {
 long classStart;
 @BeforeClass
 public void setUp() {
    classStart = System.nanoTime();
    long threadID = Thread.currentThread().getId();
    System.out.println(String.format("Class starting on
        thread %s ...", threadID));
 }
 @Test
 public void test1Time() {
    long testStart = System.nanoTime();
    long threadID = Thread.currentThread().getId();
    System.out.println(String.format("Starting test1 on
```

```
      thread %s",threadID));
    long testFinish = System.nanoTime();
    System.out.println(String.format("Test 1 took:
      %s nanoseconds", (testFinish - testStart)));
  }
  @Test
  public void test2Time() {
    long testStart = System.nanoTime();
    long threadID = Thread.currentThread().getId();
    System.out.println(String.format("Starting test2 on
      thread %s", threadID));
    long testFinish = System.nanoTime();
    System.out.println(String.format("Test 2 took:
      %s nanoseconds", testFinish - testStart));
  }
  @AfterClass
  public void tearDown() {
    long classFinish = System.nanoTime();
    System.out.println(String.format("Class took: %s
      nanoseconds", classFinish - classStart));
    System.out.println(String.format("or %s milliseconds",
      (classFinish - classStart) / 1e6));
  }
}
```

Some considerations on the previous example:

- Please note that we are also printing the thread ID where each piece of code executes. This will be important later on when we talk about parallelism.

- We are using System to print the results. This is not always the best practice; we might want to use a logging library instead as those could be ported outside of the terminal logs.

- We are using nanoTime instead of getting milliseconds directly as we are not doing much in the tests themselves, and otherwise, we might get only 0 as it is not as precise.

Keep in mind that the benchmark that you need for your application might be a bit more detailed or specific to your needs. For instance, if you run many or long tests, it might have a different effect on the system. You can also add this logging logic to your existing **build verification tests** (**BVTs**) so that the results are closer to your actual system.

It is also crucial to run the benchmark at different times of the day, on different days of the week, and—if possible—against different regions. This will bring you information about the speed of the providers in time for machine assignments.

You might need extra code to call the providers within the benchmark or to measure the results from their logs as well so that we understand the time that every step required and can reach better conclusions by achieving better comparisons.

Running this benchmark locally without parallelism will give us the following results:

### Results from benchmark.java

```
Class starting on thread 1 ...
Starting test1 on thread 1
Test 1 took: 162400 nanoseconds
Starting test2 on thread 1
Test 2 took: 230800 nanoseconds
Class took: 155514700 nanoseconds
or 155.5147 milliseconds


===================================================
Default Suite
Total tests run: 2, Passes: 2, Failures: 0, Skips: 0
===================================================
```

You can see that the thread used was the same and that the nanoseconds are different per test (and will differ altogether if we execute them again, as it might take a different amount of time for the computer to process them). Having a few executions of the benchmark could give us some average times, but we are looking for bigger differences here to have a point to select between platforms.

We now have a better understanding of how to measure the performance of our tests to understand the speeds of different providers. In the following section, we will see some tips to test appropriately in the cloud.

## Testing appropriately in the cloud

Some people might just consider the cloud as a test aid to achieve support for more devices or browsers, then proceed to use it as they would use their current system. However, the cloud offers much more than that, and it is important to use it in an appropriate manner to get the best out of it.

In this section, we are going to have a general look into how to run your tests on a cloud test provider. Then, we will discuss how parallelizing tests and orchestrating tests can be achieved in the cloud.

There are *two* possible ways to execute a test in the cloud:

*The first one* is to upload the entire test project into the agent that executes the test.

*The second one* is to use some provider's particular desired capabilities and a remote connection, such as in the following example:

## General provider example

```
driver = webdriver.Remote(providerURL, providerCapabilities)
```

Here, the `providerURL` parameter should include a username and an access key (`username:access_key`) or an API token. For Appium tests, the provider URL usually ends in `/wd/hub`.

For example, and as a comparison, let us see the provider URL of the following cloud testing providers:

## BrowserStack provider URL

```
'https://YOUR_USERNAME:YOUR_ACCESS_KEY@hub-cloud.browserstack.com/wd/hub'
```

## Sauce Labs provider URL

```
"https://SAUCE_USERNAME:SAUCE_ACCESS_KEY@ondemand.saucelabs.com/wd/hub"
```

## HeadSpin provider URL

```
"https://proxy-us-sf-3.headspin.io:7003/v0/<your_api_token>/wd/hub"
```

The code for AWS Device Farm is slightly different than the general provider example previously shown, as the URL is given by the test grid response from the Device Farm client, which depends on the account ID, project ID, and region name, rather than a fixed one or one that depends on keys or tokens:

## AWS Device Farm

```
devicefarm_client = boto3.client("devicefarm", region_name="us-west-2")
testgrid_url_response = devicefarm_client.create_test_grid_url(        projectArn="arn:aws:devicefarm:us-west-2:accountID:testgrid-project:ID", expiresInSeconds=300)
driver = webdriver.Remote(testgrid_url_response["url"],
```

```
capabilities)
```

The desired capabilities vary greatly depending on the device or browser to test and on the cloud test provider. Please check the documentation of your provider of interest to understand those.

## Parallelizing tests

As we mentioned before, an important point of using the cloud is to help with parallelizing tests. The objective is to reduce the total time spent on testing while ensuring all cases are being dealt with. Having one test per execution is the ideal scenario, keeping in mind the issues learned in *Chapter 4, The Secret Passages of the Test Pyramid – The Top of the Pyramid*.

However, the test languages or providers might have a limitation in the total number of parallel tests that can be executed. The best solution for that case is to split the tests into sections and call them in batches for each of those sections. How you distribute your tests is very important in order to achieve the biggest parallelization possible (as long as splitting them makes sense).

Most test frameworks would allow you to group the tests for parallelization, but in the extreme situation that you have more tests than can be run in parallel, we could section the code to make sure they run in parallel as separate drivers or in separate machines altogether.

Here is an example of a way you could section your test project into two so that you could call them on a manual parallelization:

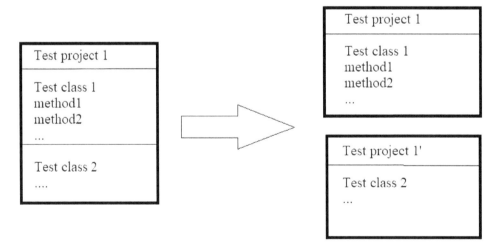

Figure 9.1: Test sectioning example 1

Balance is key—it is worth mentioning that we should keep a good balance in our test pyramid, avoid test repetition across it, use the tools from this book (including visual-based testing and test analysis), and try to keep as few tests running in devices or browsers as possible. All of these could also be used in combination (for example, executing visual tests in the cloud).

Keep in mind that parallelizing tests might result in reserving more machines, which could potentially increase the cost of the cloud provider. Furthermore, if the test takes less time to run than it takes to reserve a new machine, it might be faster to run a few tests together instead.

Let us illustrate this with an example:

| Test ID | Total time (seconds) |
| --- | --- |
| #1 | 2 |
| #2 | 4 |
| #3 | 5 |

Cloud machine retrieval and cleanup take 10 seconds.

1.  If we reserve one machine and run it all, it will take the following amount of time:

    A.  2 (test #1) + 4 (test #2) + 5 (test #3) + 10 (machine) = 21 seconds in total.

    B.  The total is therefore 21 seconds.

2.  If we reserve one machine for tests #1 and #2 and another machine for test #3, it will take the following amount of time:

    A.  2 + 4 + 10 = 16 seconds.

    B.  5 + 10 = 15 seconds.

    C.  The total will be 16 seconds, assuming both machines run at the same time.

3.  If we reserve a machine per test, it will take the following amount of time:

    A.  2 + 10 = 12 seconds.

    B.  4 + 10 = 14 seconds.

    C.  5 + 10 = 15 seconds.

    D.  The total is therefore 15 seconds (again, assuming the three machines run at the same time).

The optimal solution, in this case, is the second one: run tests #1 and #2 in one machine and test #3 in another machine. If reserving machines is too expensive, then we might put up with the longest test of solution 1. Solution 3 takes almost the same time as solution 2, but we needed an extra resource or machine reservation, which could involve extra costs. We could achieve this sectioning across tests by grouping the tests and running per group or by sectioning the tests as we saw before (do that only if your testing framework for some reason does not allow for any sort of grouping; it is not advisable otherwise).

For example, if we want to group our tests with TestNG, in the previous class (`benchmark.java`), we will add the following annotations after `@Test` and on top of our test method code:

## Grouping in TestNG

```
@Test(groups = { "group1" })
```

Here, `group1` is the name of the group we want to assign to said test. In our case, `test#1` and `test#2` could be `group1` and `test#3` in `group2`:

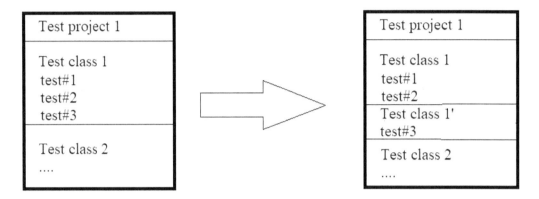

Figure 9.2: Test sectioning example 2 – Grouping tests to run them conveniently

This is a simple example, and generally, we do not need to go into such a deep level of detail, especially if we have enough bandwidth or machines in the purchased cloud resources. However, this goes to show how important it is to benchmark tests and analyze them to achieve the best speed using the fewest resources.

That said, all the analyzing, grouping, or splitting code will add up to the test creation time, so make sure it is worth it to spend time on it—that is, if the time invested will be reduced in the long term by the time or costs saved in execution.

Following our first example for the chapter, we could run parallelization using TestNG: the tests can be parallelized by methods, tests, classes, or instances. The default number of threads that could run simultaneously for TestNG is 5. We can set the thread count in the test annotations (when we write @ before the method or class) or in the XML test configuration file. We can also parallelize the tests at the data-provider level in a similar fashion through test annotations.

Note that the following files can be located within the `Benchmark` folder in our GitHub repository.

Let us see what the `testng.xml` file would look like for parallelizing the first example of the chapter by `test`:

### testng.xml

```
<?xml version="1.0" encoding="UTF-8"?>
<!DOCTYPE suite SYSTEM "https://testng.org/testng-1.0.dtd" >
```

```
<suite name="parallel suite" verbose="1" >
  <test name="AllTests" parallel="methods" >
    <packages>
      <package name="testingTimeMachines.chapter9" />
    </packages>
  </test>
</suite>
```

And running it again, the results now are as follows (check specifically the thread ID values):

```
Class starting on thread 10 ...
Starting test2 on thread 11
Starting test1 on thread 10
Test 2 took: 281200 nanoseconds
Test 1 took: 1005300 nanoseconds
Class took: 20345600 nanoseconds
or 20.3456 milliseconds


===================================================
parallel suite
Total tests run: 2, Passes: 2, Failures: 0, Skips: 0
===================================================
```

Finally, instead of using packages to define the tests, we could use classes or groups, as in the following piece of code:

## testngByGroup.xml

```
<?xml version="1.0" encoding="UTF-8"?>
<!DOCTYPE suite SYSTEM "https://testng.org/testng-1.0.dtd" >
<suite name="group parallel suite" verbose="1" >
  <test name="Group1 tests" parallel="methods" >
    <groups>
        <run>
            <include name = "group1" />
        </run>
    </groups>
    <classes>
```

```
            <class name="testingTimeMachines.chapter9.Benchmark" />
        </classes>
    </test>
    <test name="Group2 tests" parallel="methods" >
        <groups>
            <run>
                <include name = "group2" />
            </run>
        </groups>
            <classes>
        <class name="testingTimeMachines.chapter9.Benchmark" />
        </classes>
    </test>
    </suite>
```

In this case, the results are not very different as we did not benefit from the grouping itself.

In order to test the code of this section, you could use your favorite Java IDE. However, if you happen to want to run the TestNG tests of this section from your terminal (given you have Java installed and TestNG downloaded in the `lib` folder of the project), you could use this command:

```
java -cp (path to the lib folder where testng is)\lib\*;(path
to the bin folder)\bin org.testng.TestNG testngByGroup.xml
```

There is no "one-size-fits-all" solution; it is up to you to figure out the best values for your system's optimal parallelization, and this book hopes to give you the resources and skills you need to achieve this.

## Orchestrating tests

In *Chapter 4, The Secret Passages of the Test Pyramid – The Top of the Pyramid*, we saw how important it is to orchestrate tests, users, and data. The examples that we saw were taking place inside of our company's network and with full access to the resources. However, once we use cloud testing, we need to send the data back and forth to the provider's network.

Orchestrating tests in the cloud has many variants, and it depends on what your tests need to do and how you need to orchestrate among them.

> **Note**
> Do not try to orchestrate the run of dependent tests in the cloud! Keep your tests independent.

Instead, use orchestration to ensure the tests do not collide with each other by using different users or data instead, or if tests need two different users to communicate.

Similarly to what we covered in *Chapter 4, The Secret Passages of the Test Pyramid – The Top of the Pyramid*, we can have more than one driver and orchestrate between them. Following that example, in the next piece of code, we test some chat functionality that orchestrates between two users. Note that the actual implementation will depend on the cloud provider's URL and capabilities:

## orchestrationExample.py

```
from selenium import webdriver
import unittest
''' define your providerURL and providerCapabilities (you will
need an account) '''
driver1 = webdriver.Remote(providerURL, providerCapabilities)
# note that this connection might take longer than a local one
driver1.get("chaturl")
# Assuming login done for user 1, in some way or having
# different URLs for different chats
driver2 = webdriver.Remote(providerURL, providerCapabilities)
# Assuming login done for user 2
driver1.find_element_by_id("textBox").send_keys("hello user 2")
assertTrue(driver2.find_element_by_id("chatBox").getText().
contains("hello user 2"))
driver1.quit()
driver2.quit()
```

The key here is that the connection and communication with the remote executor might take longer than connecting them locally or even within your network.

If your cloud provider needs to upload the entire test for it to run, rather than a remote connection, you will need a different logic to handle the test orchestration. For example, you could have two test files and wait for events to happen in each of the tests rather than having it all in the same test. This is not an ideal situation to be in; keep in mind that the tests would now be dependent on each other being executed. Even if you were to parallelize them as a group, they could be prone to failures due to the time they get executed being non-deterministic. If you face this issue, you should search for solutions within the test provider to try to guarantee the execution of both tests within a particular timeframe.

In this section, we have seen how to parallelize tests and how to orchestrate them, sending resources outside of our network. However, these resources might be sensitive data, which would be insecure to send via the internet. In the next section, we will discuss the dangers of the cloud, such as this.

# Dangers of the cloud

In *step 1* of the *What exactly is the cloud, and how can it help you?* section's list for selecting a cloud provider, security is mentioned as one of the features you might need, especially when you are dealing with sensitive data.

Providing a secure system is in the interest of all cloud providers, but sometimes there might be some limitations due to the nature of the cloud. Not only do you have to trust that the company itself will not take advantage of your system, but also that it will keep it secure from external malicious individuals.

Missing issues due to the tests being executed in simulated devices rather than real ones, is also a danger of the cloud, so make sure you confirm what case it is going to be and cover yourself from it.

As we will see in a bit, one more danger could be that the provider does not clean up the application and tests after their execution, and since devices are shared across clients, some other client might have access to your data.

Since your tests are performed at a different location, apps and/or tests have to travel to that location, which makes them vulnerable. For that reason, if your system needs strong privacy and security, it is advisable to invest in your own device farm, especially if you don't need to test in different global locations of very specific devices.

## Considerations for creating your own device farm

There are many considerations that you need to have for your own device farm. Firstly, you need to design your system well, considering which devices you will support, where they will be physically located, what sort of security you need around the connections and the system, as well as physical wires and connections.

You also need your devices to be connected to a computer that would perform the orchestration among them. Since this computer is not going to be doing expensive operations, it could be a simple one (such as Raspberry Pi). However, it might need to comply with your company's policies, and that would mean that it might need to have some specific software installed on it. This computer will also need to support handling a number of devices, so you might need to scale to several of them orchestrated by another one.

You might also need other resources—for instance, a database to keep track of the statuses of the device, a system to alert of issues, and the logic to perform smart orchestration among them.

Overall, creating your device farm could incur additional costs not only in the devices purchased and the coding involved in the handling of devices and security but also in maintenance, as if anything goes wrong you would need someone physically near the devices to fix the issue personally. Finally, it is recommendable that you perform some tests outside of your own network to guarantee that the system behaves the same and does not take for granted connections that the user might not have.

On the other hand, you will have full control over the devices and the particular models you require, metrics such as peak utilization hours, access, and better privacy for your shared data.

## Using test cloud providers securely

Test cloud providers often offer some security solutions such as local testing, encryption, or tunnels around the connection to make it safe to perform the tests. Make sure this is well done and tested and that it is secure enough for your system before engaging with it blindly. If it happens not to be as secure as they claimed, or you did not understand fully the risks of using their services, the repercussions might be damaging to your product and the company's reputation.

Another security concern arises from the fact that the devices are shared. You should always make sure your data is cleaned up before the next user makes use of the same device to avoid leaks. A plus would be for them to have devices owned just by you, especially for sensitive tests, but that is generally much more expensive, and it could be almost as costly as handling your own device from within your company without using the cloud.

Before taking testing to further devices, it is likely that you already have done an amount of testing locally. If you decide to move to a cloud solution, it is advisable that you keep some local tests running now and then and/or use your local ones for the most sensitive tests. You could even have a combination of external cloud farm and internal device farm.

As usual, the right balance is key.

A couple of additional tips are to make sure that your new features are ready for exposure by the time you use the cloud services and that you don't use sensitive data, change passwords frequently, use cryptography and code obfuscation and authentication tokens, and do not expose internal resources such as databases. If possible, check with a security expert before engaging with cloud testing and keep them in the loop of conversations for further guarantees.

## Testing applications in the cloud

The other side of cloud testing is when you have applications that, fully or partially, reside in the cloud. This is the case for serverless applications.

> **Serverless application**
>
> An application in which part of it is handled in the cloud—that is, there is a provider that will take care of maintenance and scalability of the (generally back-end) systems required for the application to run.

You might think that the difference between testing fully hosted applications and partially hosted applications is that you can skip some of the tests that generally would take place in those areas. However, while you could skip extensive testing for those areas, it is still important to test the stability of the system and assign contracts and verifications across it to make sure everything works as expected.

In *Chapter 3, The Secret Passages of the Test Pyramid – The Middle of the Pyramid*, we covered tests such as the ones just mentioned that could be applied to cloud applications as well, especially validating integrations and contracts between each of the parts.

As another example, we might trust that our cloud provider handles the database in the right way (or we could double-check this for our own security with some basic verification tests). However, additionally, it could make sense that we still want to add some tests of our own to verify something specific about the data we add to the database.

Besides all of this testing, it is still important to set up benchmarks, measure the performance of the different components of the application, and keep metrics for everything to make sure all the components work well and connect appropriately. The benefit of keeping track of it all is that we should be able to easily substitute a component on a cloud provider for another one on a different cloud provider, in case we conclude it is not up to our standards or find a better deal. It could also be a good idea to have some facade methods to call the provider, so that if we need to change the calls, we just change the facade rather than having to change each call across all of our code.

In the next section, we will see other ways of thinking about the cloud that could benefit our test architecture and automation.

## Thinking out of the box about the cloud

There could be many other advantages to using cloud computing other than speeding up testing and deployments. For example, we already covered this in *Chapter 6, Continuous Testing – CI/CD and Other DevOps Concepts You Should Know*. CI (and/or testing) could be performed and installed in the cloud by using runners.

---

**Runner**

In CI/CD, a runner is an agent (a program) available in one or more machines that can execute jobs (pieces of code).

---

Runners can be local or live in the cloud. For this chapter, we are considering mostly the ones in the cloud.

Runners can perform several actions, including setting up a Dockerfile to prepare the runner with the required programs to install and run the app. They can also run the test code, which will call the cloud provider if we are using one. That would be a cloud that is talking to another cloud, about potentially yet another cloud (if your app is serverless):

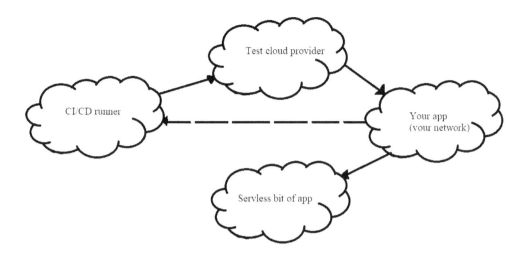

Figure 9.3: Example of a cloud architecture

Runners could be utilized to run multiple programs, allowing us to automate different steps of the CI/CD ecosystem. Among these, we could utilize the runners to align with other programs or tools—for example, a reporting tool.

A good example of automation that can happen in the cloud directly against your repository is the case of **GitHub Actions** (see *[5]* and *[6]* in the *Further reading* section).

> **GitHub Actions**
>
> A platform that utilizes YAML instructions to automate actions in your GitHub deployment pipeline, including actions from within the GitHub repository.

In *Chapter 6, Continuous Testing – CI/CD and Other DevOps Concepts You Should Know*, we talked about GitHub Actions and YAML files. In this chapter, we will go beyond CI/CD and discuss GitHub Actions' functionality for repository automation.

The GitHub Actions YAML configuration files are generally stored under `.github/workflows` in your repository. Let us see a basic example here:

**.github/workflows/easy_action.yml**

```
name: EasyAction
on:
  issues:
    types: [opened]
```

```
jobs:
  newJob:
    runs-on: ubuntu-latest
    permissions:
      issues: write
    steps:
    - uses: actions/first-interaction@v1
    - with:
        repo-token: ${{ secrets.GITHUB_TOKEN }}
        issue-message: 'Easy action triggered'
```

Note that this is sensitive to indentation and blocks of commands; make sure you use the code from GitHub and save it in the appropriate location rather than copying it directly from here.

As you can see, this is very similar to what we discussed in *Chapter 6, Continuous Testing – CI/CD and Other DevOps Concepts You Should Know*. In this example, every time an issue is opened (on code block), an action (`actions/first-interaction@v1`) is executed using the indicated repository token and message. This particular action adds the indicated message the first time an issue is opened in the repository.

The code with the logic of this particular action is a program written in the TypeScript language, located here: `https://github.com/actions/first-interaction/blob/main/src/main.ts`.

A benefit of having these actions within the repository is that you could use the repository's secret variables to save environmental values, saving time to set those in the runners directly and with easier and more private access, directly from within the `actions` YAML file.

This is done as follows:

---

### .github/workflows/github_actions_workflow_name.yml

```
. . .
${{ secrets.YOUR_SECRET }}
. . .
```

As you might have noticed, a similar line appears in the `easy_action.yml` example, retrieving a secret called `GITHUB_TOKEN`. There are some predefined secrets in the repository, such as this GitHub token, but we could define others—for example, test users or passwords or any other sensitive data. These variables can be configured from GitHub's web interface at repository, environment, or organization levels, allowing sharing if need be.

GitHub actions are created by the open source community, and many tools have their own actions that integrate with the GitHub repository.

For instance, BrowserStack and Sauce Labs both have their own GitHub actions that allow them to open a tunnel between their services and a runner behind a firewall and/or proxy. At the time of the writing of this book, there is also an AWS Device Farm GitHub action that allows your repository to interact with AWS Device Farm. These examples can be found easily with an internet search, and they could change by the time you are reading this, so we are not going to showcase them in this book.

There are multiple other examples of GitHub actions that could be useful for your application, depending on which tool you use. For example, you can run and analyze tests in Cypress, connect with reporting tools, send alerts about failures in different formats, connect with issue repositories, check for coverage, and analyze and visualize web performance.

We highly encourage you to practice trying to build actions yourself. A good starting point would be taking an existing action and making some changes. For example, for practice, you could clone the repository of the first action that we linked before and try to make it so that it will add a message for every opened issue (not only the first one). You could also play with the YAML file, changing when the actions are performed; for example, when the issue is closed, mentioned, locked, ready for review, and so on...

## When not to use the cloud...

While these actions can be of use for many automations, we should also keep in mind aspects such as pricing, security, and execution time, especially if we use shared workers rather than configuring our own servers.

Lastly, keep in mind that not all types of testing can be done in this way. Load testing is a good example of a test that does not make sense to do in this way, especially if we are using runners in the cloud. A test such as this would likely be blocked by the server where it gets executed as it might be considered a **denial-of-service** (**DoS**) attack by it. We should not try to skip this failure as it might open a security gap for someone malicious trying to access our servers. Besides, if the runners are shared in the cloud, the test may not yield the right results as it might happen that the runner is balancing the execution of different tests from different clients with ours.

Unless it's for browser or device performance measurements, it is best that we execute our performance, load, and even some chaos testing in a controlled environment. Then, we can assume that the results are from best-case scenarios and monitor our resources in production with solutions such as the Elastic Stack.

As we progress and build more tools in the cloud, it is likely that we will come across new functionality and capabilities for automation that could be of use to us to improve the quality of our applications more safely.

If you want to achieve more while saving on maintenance, think about which other parts of what you are currently working on could be done "in the cloud" and what benefit could it bring for your product to have them there.

## Summary

In this chapter, we have reviewed several ways of seeing and using the cloud in relation to testing and test architecture. We started the chapter with some definitions, followed by how to pick the right cloud provider. Then, we looked into how to test appropriately and the potential dangers of using a cloud provider for testing. We then looked into the opposite side of the cloud: testing an application in the cloud. Finally, we reviewed other ways we could use the cloud to our benefit for testing.

In the next chapter, we will take a trip into virtual, augmented, and cross realities.

## Further reading

- [1] AWS Device Farm documentation: `https://aws.amazon.com/device-farm/`
- [2] BrowserStack documentation: `https://www.browserstack.com/`
- [3] HeadSpin documentation: `https://www.headspin.io/`
- [4] Sauce Labs documentation: `https://saucelabs.com/`
- [5] GitHub Actions introduction: `https://github.com/features/actions`
- [6] GitHub Actions documentation: `https://docs.github.com/en/actions`

Make sure you continue pursuing knowledge of other topics related to the cloud: different providers, cloud testing (testing the cloud), and testing in the cloud (testing using the cloud).

# 10
# Traveling Across Realities

XR, also known as extended or cross reality, is a concept that includes **Virtual Reality** (**VR**) and **Augmented Reality** (**AR**) among other similar technologies. In this chapter, we will cover different definitions so that even if you are new to XR technologies, you can get a good understanding of how everything is connected and what could potentially go wrong, providing you with the knowledge to test your XR projects. Even if you are currently not working on an XR project, as computers' components become more powerful, and new tools and algorithms come out that enable easier-to-build and better-performing XR applications, the importance of knowing these types of applications and how to test them also increases.

In this chapter, we will cover the following topics:

- Getting started with XR
- VR and testing
- AR and testing
- Finding XR bugs and testing in XR
- Quaternions and why they are important in XR
- Tools for XR development
- Tools for XR testing
- The metaverse

By the end of this chapter, you will have a clear idea about how to develop and test XR applications so that you can easily build on top of them and take your XR quality to the next level.

# Technical requirements

XR is an advanced topic that not everyone uses in their day-to-day jobs. We hope you can still enjoy this chapter without the knowledge of cross-reality (VR, AR...) development, although you will get the most out of this chapter if you are familiar with some of these XR topics and maybe have even done some programming for XR. However, there should be enough information about the topics for beginners to be able to understand them, too.

This chapter could also serve as a fun introduction to XR, to get comfortable with it and allow you to learn more on your own. As with the other chapters, the code related to these examples is presented in our GitHub repository: `https://github.com/PacktPublishing/How-to-Test-a-Time-Machine/tree/main/Chapter10`.

Note, we have also uploaded an XR sample app in there that we will use for the testing examples.

# Getting started with XR

The concept of XR came about as a way of referring to a combination of technologies such as AR, VR, and **mixed reality** (**MR**), among others. If you are already involved with XR, you might want to skip this section as it is quite introductory, unless you want to review these concepts.

We will explore each of the mentioned terms in the following definitions. All of these technologies involve having something that is not real, and that has been created by a piece of software.

The first *non-real thing* we could have is an object that is displayed in our real world, such as a sort of hologram or a display (as, it could be a screen or a projector).

> **Augmented reality (AR)**
>
> The user experiences virtually created objects appearing in the real world, or real objects are enhanced virtually. This is usually achieved by providing the objects within some display (such as your mobile phone) or projecting them. Although most current technologies would be visual, other senses are also included as part of the AR definition.

Many applications use AR, for example, games such as **Pokémon Go** *[1]* and the applications that preview the color of paint on walls, how clothes would look on you, or how a piece of furniture would fit inside a living space. Additionally, some museums offer AR experiences in their displays.

If we were to surround ourselves with virtual objects to the point that we abstract ourselves from the physical world, we would have our second *non-real thing*, an entire world!

> **Virtual reality (VR)**
>
> Here, the user experiences a virtual or artificial world. As before, the experience might include several different senses. This is, usually (and currently), achieved with the incorporation of a head-mounted device that displays the world to the user.

Besides games, there are VR applications that help users become immersed in different situations in order to create empathy or recover from certain traumas.

The final *non-real thing* that we could have is a mix of the previous two.

> **Mixed reality (MR)**
>
> Here, the user experiences the real world and the virtual world interacting together. For example, having a door that the user can step in to go to the virtual world and back, and throw virtual objects to the real world, with these objects being able to bounce against real-world ones. Generally, you would need the same type of devices for MR as for VR.

While most people might think about XR as a fun concept, many XR applications are very serious and allow humanity to achieve advances in every field we could possibly think of. For example, we could have an XR application to help a doctor perform an operation, for training in different scenarios, and more.

Whatever the application is built for, fun, learning, research, or business, they all have one thing in common: they are tremendously challenging to test. Therefore, before embarking on such an adventure, it would be good if we gain some initial practice and experience.

## How to get experience in XR

Since XR applications are still gaining momentum, it is likely that you have not had the chance to interact with them but you would like to learn more. As with other topics in this book, there is a lot of material for developing XR applications, including books, online courses, in-person courses, and open source projects.

There are many free courses for XR development if you do not care for a certificate or a degree, and some learning platforms might allow you free access to their content for which you can pay a fee later on to earn that degree.

If you are more of a *hands-on* person, you might like to try some apps on an actual device. The cheapest one, for any XR applications, would be your phone, as you are likely to own one already. Check for AR, VR, or XR apps or games in your phone's store. If you really like the XR experience and would like to purchase a more expensive device, the device's seller or brand store would likely point you to the best place to check out some apps.

As mentioned before, checking the source code for some of these apps can help you gain a better understanding of the technology.

Understanding how apps are developed is essential to understanding how to better test them and the needs of the development team. But seeing them from the users' perspective is also quite important. Therefore, even if you don't have access to an XR device or app and/or are unable to understand the code related to them, you can still find out how users use the apps. For example, for games, there are several playthrough videos of XR games available online. Additionally, there are some videos of XR experiences that failed for the user, which is important to acknowledge and visualize for testing purposes. In some cities, there are events or conferences with XR content where you can experience it or watch the experiences of others.

The tools that we will review in the last two sections of this chapter will also help you gain more knowledge about the XR space, and most of them provide emulators with which you can check how your virtual assets look and interact with the created environments and users. They come with plenty of documentation that you can check out even before installing them.

However, before getting into the specifics of XR development and testing, let us learn more about VR and AR.

## VR and testing

In this section, we will see cases and stories of VR and testing. Let us review the concept of VR and see how applications for it are created.

### VR development

As we saw in the *Getting started with XR* section, VR is about creating realistic artificial worlds for the user to view or interact with.

Generally, VR is achieved by the use of a headset that the users would place upon their heads. However, there are other ways to achieve this. For example, there are simpler versions that do not cover the entire head but work as sort of glasses, which do cover the eyes from external lights.

To cover the part of the interaction with the virtual world, the simplest devices have a point in the middle of the screen (called *reticule*) that will cause an interaction if it stays on an object for some time. Some other devices add hand-holding controls that allow the user to interact with the virtual world using their hands. More advanced devices can provide other hardware and ways of interacting with the virtual world, using other parts of the human body to interact with the application, for example, using your ears and interacting by sound.

As XR systems become more developed and more affordable to build, buy, and distribute, we will probably see more complete and accurate devices and environment experiences other than handsets and headsets, such as moving platforms or smelling sensations that we should also consider for our testing. Other systems add the tracking of the user's body parts, such as the mouth, the eyes, or even the heart.

In order to track the movement of the head of the user, the devices can use a sensor to determine their orientation based on Earth's gravity (called a gyroscope). Other devices have the capability of detecting your movement within a room as well, using many different technologies for that.

Some of the devices are connected to a system such as a PC, console, or phone. Others are wireless and standalone. There are some applications that run on a web browser and others that are specifically built for certain platforms.

All these differences will have their own uniqueness in the way we test them. Let us take a look at the ways we could test a VR application.

## Testing in VR

In order to consider how to test an application in VR, recall *Chapter 2, The Secret Passages of the Test Pyramid – The Base of the Pyramid* to *Chapter 4, The Secret Passages of the Test Pyramid – The Top of the Pyramid*, where we examined the test pyramid and the different types of testing. Similar types of testing could be applied to a VR application, putting more weight on one type or another depending on the application itself.

However, there are some particularities that make the VR environment more challenging for testing than other applications. For these apps, it is especially important that we adopt the mindset of the user, and make sure we understand the differences of human beings that could make something feel uncomfortable or difficult for people different from us – for example, just a user's height can make a difference in terms of how comfortable it is to reach for objects. Accessibility testing is especially critical for these apps.

Let us see some tests that are specific to VR.

### Spatial sound

Since the world surrounds the user, in order to create realistic experiences, it is common for developers to integrate sound to guide the user to look at where the action is happening or help them find hidden objects.

### Physical surroundings

In VR, we are immersed in a virtual world, so we might not be aware of the actual world around us. We should make sure the user is safe there too, providing ways of disconnecting easily when an emergency occurs and reminding the user to be careful with objects nearby.

### Motion sickness

In VR, people can experience similar effects that some people experience in boats at sea, including dizziness. This happens because the movements your eyes are seeing do not match your actual movement. This does not affect people in the same way, so testing it could be a little bit challenging.

### Closeness to objects and other players

Some people might feel uncomfortable when objects, other players, or characters within an app enter their personal space.

### Objects that cannot be reached

The size of the surroundings and the movement of objects might make it easy for an object to land out of the reach of a user, making it impossible for them to proceed with its use.

### Loading and other performance issues

VR components are expensive to generate for a computer, and issues with performance or loading, such as the frame rate per second decreasing at important times, are common and should be carefully measured.

### Issues with lighting, textures, and objects

Even if the world is virtual, the objective is to immerse the user in it, so we generally would want that things should look as realistic as possible. The stories could play with the size of the objects to make it seem as though the user is small or big within the world, but when effects like that are not happening by purpose, then we could run into some issues.

A badly imported texture could break the immersion of a user or make some users unable to see and, therefore, interact with a particular object. The same could be said for lights that do not match their logical paths. These particularities could be tremendously difficult to automate for testing, so most of the time, they require manual checks.

### Object interaction

Collision, visibility, bouncing, and such need to be taken into account and reviewed.

As we have just seen, there are different types of VR devices and platforms, which results in different tests required for each of them. For example, for devices with handsets, the functionality of the different buttons on the different parts of the VR application should be considered. Keep your hardware in mind when you enroll in the testing of a VR application.

In the next section, we will review the particularities of AR and its testing.

## AR and testing

AR has a lot in common with VR, starting with development systems. Therefore, the testing of the two platforms is very similar. However, AR simplifies anything in relation to the hardware that VR provides. Therefore, with AR, motion sickness does not happen as frequently, as users can naturally be more aware of their surroundings. However, there are occasions in which they are not, such as when they are looking at their phone instead of their path, which could be hard for us to avoid, although we could provide the users with warnings to remind them to be aware of their surrounding and to not use the application if it is unsafe to do so.

On the other hand, objects that cannot be interacted with could be a common issue in both AR and VR; for example, if the object appears inside a physical object or in a place that is unreachable (such as in the middle of a swimming pool, a road, or if it is too high for the user to reach).

Another thing AR has in common with VR is loading and performance issues. However, AR systems generally load fewer objects than a virtual world could load, so they are usually less heavy on these sorts of issues.

A particularity for AR systems that VR does not have is that in AR, virtual objects should be visible against the different contexts and surfaces of the real world. The test set for each particular system should include situations in which objects might blend with the environment.

Other types of testing, such as those we mentioned in *Chapter 2, The Secret Passages of the Test Pyramid – The Base of the Pyramid, Chapter 3, The Secret Passages of the Test Pyramid – The Middle of the Pyramid*, and *Chapter 4, The Secret Passages of the Test Pyramid – The Top of the Pyramid*, covering the test pyramid, should equally be taken into account, and we will review them in the next section.

# Finding XR bugs and testing in XR

In the previous two sections, we looked at specific tests in AR and VR. In this section, we will review the more general testing that can be done in XR, although the testing that we cover for VR and AR should also be taken into account for any XR application. Let us perform the exercise of reviewing the test pyramid and mention some particular tests we should perform for XR.

## Base layer

As with any other application, the units of code for an XR application should be carefully tested, especially when measuring code coverage and code optimization.

Unit testing in XR should focus on three main aspects of an XR application:

- **The code**: This is just like any other application.
- **The object interaction**: This includes collisions, bouncing, removing, creating, or changing objects... Make sure all of these interactions are properly put to the test during unit testing (or, if not possible, later on), whether the objects are interacting with other in the physical or virtual worlds.
- **The scenes**: The environment of the XR application should also be carefully put to test. Some applications will also change scenes or perform changes on them through the code, so we need to make sure they are performed correctly.This includes possible transitions between virtual and physical worlds, if they occur.

Some parts of the application could be autogenerated as some development platforms allow for the usage of the UI to develop parts of the application.

One of the specific things we could be looking for in code reviews is the use of Euler vectors instead of quaternions. We will talk about this a bit further on in this chapter, in the section *Quaternions and why they are important in XR*

## Middle layer

Testing your application in the different systems and platforms that it is being developed for is crucial in XR. For this, you should be perfectly aware of what systems the application supports. Additionally, we should make sure that the hardware responds adequately to the inputs given during its usage.

Some XR applications also allow for communication with other systems, such as in the case of a game that compares your results with other players around the world or retrieves content in a dynamic way. This could be done through API calls, so make sure you test those too.

Sometimes, the application will save its results on a database, which means we should take care to test our database too.

We will not cover other types of testing within the middle of the pyramid, since the specific testing might vary in each application, but make sure you check all those related to your application and think carefully about whether there is anything special that you can do with them in XR.

## Top layer

Accessibility testing is crucial for XR apps. We are not all built in the same way, so we should make sure that our systems can be enjoyed and utilized by as many users as possible. This could be achieved by allowing for different configuration options that can make the application more friendly to different people.

As mentioned in the VR section, XR applications are computationally expensive to execute. The more realistic we try to make an environment, the more objects, materials, textures, and lights we need to include, and that results in greater difficulty for the computer to load it all We should take into account all the different hardware that we support, and make sure the system runs fine for each of them (for example, lowering the quality for older systems and allowing the user to configure this). This includes the different platforms supported (for example, a computer versus a phone).

Some of the key performance items to check for XR applications are GPU and CPU usage, rendering performance, audio, physics, memory usage and leaks, and battery utilization. If the application uses the network, network impact could be another point to check, along with crash reports. If the application is online, when possible, make sure your system has alarms in place for tracking the issues and for a quick turnaround if we find them.

For multiplayer applications, we should ensure the number of users that can connect at a time without issues.

Security testing is always important in any application. As the platforms evolve, new attacks could arise too; for example, virtual property robbery.

Besides all of this, normal functional testing should also be performed. Generally, this is done in a manual manner, although some recording and playback tools might be developed to assist with this effort.

## Other testing

We could consider usability testing, which we mentioned earlier, such as making sure that objects can be seen and reached comfortably (by any user) and appear realistic.

Motion sickness (covered for VR) could also be a problem in XR. There could be additional issues, for instance, when you jump from a static physical world into a moving virtual one.

An XR environment consists of everything that surrounds the user, that is, 360 degrees of rotation in every coordinate and combination. That means we need to do extensive testing of everything that might happen around the user at any point. Therefore, the control of the camera could be automated.

Additionally, we should cover localization testing if we include text or audio that has been translated into several languages for the worldwide delivery of our app or game.

In the case of MR applications, we could use the advantage of understanding our physical world to avoid issues with it, as mentioned in the VR and AR sections. For example, we can make sure that the sounds outside the virtual world are heard by the user so that if someone is calling to them, or some alarms are going off, they can be aware of them. We should also ensure that the virtual world integrates with the physical objects so that the user is aware of them. The remaining problem is whether they think a VR object is a real one or that an external noise is a virtual one, but that could somehow be highlighted during the experience.

## Testing with XR

XR applications serve multiple purposes, they are not only useful for immersive gaming experiences. There are several tools that could help users learn how to perform tasks that might otherwise be expensive or difficult to train on in the real world. I would love to see a tool that teaches testing over XR, and I am sure much more is yet to come.

In the next section, we will talk about an effect that occurs with objects in 3D that could cause some of the aforementioned issues.

# Quaternions and why they are important in XR

In *Chapter 7, Mathematics and Algorithms in Testing*, we saw the importance of specific mathematical concepts in software and testing. Quaternions are one such example.

In XR, dealing with **three-dimensional** (**3D**) objects is crucial for creating accurate experiences. Understanding how objects are represented can help us understand the problems 3D objects could create, and ultimately, this will help us find any issues and defects in our XR applications. Besides, this concept is also important in other fields such as computer vision.

There are different ways of representing such objects and their position in 3D space. The most well-known way is by selecting a point in our space (which could be the middle of the space or a corner, for example) and indicating what distance in each direction the middle (or equivalent) of our object is from that point. This is called the **Euler vector**.

---

**Euler vector**

A vector with three components that indicates a point in 3D space.

---

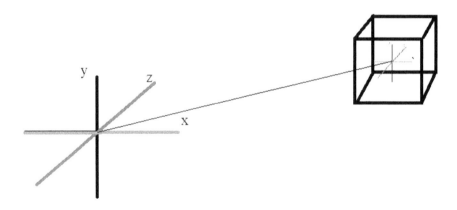

Figure 10.1: Representing an object in 3D space

Usually, Euler vectors are quite intuitive. If you need the object to move horizontally, you can make a variation over the value of the *x* axis. If you need it to be rotated horizontally, you can rotate the *x* value by an angle with that *x* axis fixed.

One of the problems with Euler vectors is that the results of two rotations might vary depending on which one of them has been done first since each coordinate rotates based on the other two. This could complicate the reproduction of issues or the reversion of the state of a particular object.

Another key problem is when two axes happen to align on a rotation. Imagine, in *Figure 10.1*, that we rotate the box in such a way that the *y* and *x* lines end up aligned in the same place. If this happens, we will lose one degree of rotation, so when we rotate *y* and *z*, the effects will be the same. This is called **Gimbal lock** (see point *[2]* of the *Further reading* section).

> **Gimbal**
>
> A pivoted support that allows an object to rotate on an axis.

Generally, there are different gimbals inside each other to allow for all the rotations. Gyroscopes, which are key to understanding the position of the head of the user in XR applications, use gimbals to function. In the case mentioned earlier, one of the gimbals gets locked with respect to another.

> **Gimbal lock**
>
> An effect that occurs when an object loses one degree of rotation.

Gimbal lock is so dangerous that even the Apollo missions had to be careful about it *[3]* since, in space, the lack of gravity results in the need for 3D measurements. This is something that is included in the navigation guidance of NASA, and we can even see it in the source code of Apollo 11 (see point *[4]* of the *Further reading* section).

Therefore, in order to avoid such issues, we need to add a redundant gimbal, or in our case, we need a different way of indicating a point in 3D space.

> **Quaternion**
>
> A vector with four components to indicate a point in 4D space.

This representation is much harder to visualize and understand. Instead of using quaternions over a 4D space, we use this representation in 3D space for rotation, to encode the angle with the point in the space and avoid the gimbal lock effect. We could think about it as rotating two entire planes at once (as opposed to Euler vectors that will only rotate one plane).

Even though the visualization of a 4D space might be harder, the mathematics applied to quaternions is generally easier than with 3D vectors, as you always get the same results for the same rotations. This is particularly interesting when we automate 3D interactions, as we can *undo* them easily, rather than having to reset all the states of the objects in the environment. It is also convenient to be able to reproduce bugs in a reliable manner.

Calculating rotations might be less intuitive, as the angles are half the expected amount for a 3D Euler rotation, so you will need to apply the rotation twice to obtain the desired result for that angle. This is called double cover.

> **Double cover (for quaternions)**
>
> This refers to a two-to-one mapping between the planes. Each rotation is represented in two ways: one applied on the axis and the other, a negative one over that axis in the opposite direction.

Understanding these concepts might take a while, and depending on what type of learner you are, it might be best to apply mathematical formulas or try to visualize them. The details of each are beyond the scope of this book, but feel free to deepen your knowledge by learning more about them.

The concepts learned in this section will, at the very least, help you understand when issues arise, and they will help you make a case arguing for the use of quaternions rather than Euler vectors whenever possible.

In the next section, we will review some tools that you can use to create applications in XR. Each of them has its advantages and inconveniences, and you should carefully research your application's needs before using them.

# Tools for XR development

In order to get familiar with the XR environment, a good idea is to try to create an XR application and check the type of testing that could be done on it. In this section, rather than looking at specific XR test tools, we will look at some XR development tools and talk about their test capabilities.

The intention of this section is not to do an exhaustive analysis and comparison for each of them, but a brief summary so that you are familiar with them and can consider a trial if you have not used them as of yet. As much as we would like to have an easy (but likely biased) answer to *which one to pick*, you should do more research about what tool would work best for your company, including development language, support, cost, performance, and any other requirements, and make sure you do a few trials before deciding on one.

One important thing to note for all of these tools is that XR applications have high performance requirements, so make sure you pick a tool that can run well on your computer, and if you need a device, make sure your computer is also compatible with that device. If you are looking to purchase a new computer, and are considering using it for XR development, make sure you take these requirements into account too.

Before starting to check the different tools, it would be good to familiarize ourselves with some typical nomenclature that the tools use, although each of them might have its own naming. For more information, feel free to check the tool documentation of each (which will be linked in their respective sections). If you are familiar with them, feel free to skip the definitions.

# XR definitions

Here is a list of some concepts that are important for XR applications:

- **CPU – central processing unit**: This is the main electronic circuit (processor) that executes the instructions of a computer program.

- **GPU – graphics processing unit**: An electronic circuit designed to perform rapid specific mathematical manipulations, primarily for rendering images, freeing the CPU from those costly operations. In XR, having powerful GPUs is an advantage. We should also check how well our app is using GPUs.

- **Assets**: The available components that are part of the library of the XR application, including textures, 3D objects, images, and scripts.

- **Game object, sprites**: The units of components of a game or XR application that are loaded in the scene.

- **Scene – Level – View – Map**: A gathering of elements or components that comprise the game or application in a particular area.

- **Camera**: We are familiar with cameras in real life, and in the 3D world, this concept is important as it will be linked to what the user is seeing at any given point. For some apps, the player or actor could be a game object within the camera, or they could be a separate instant, with the camera staying static during the experience, or moving alongside the player, or we could have more than one camera to provide different, switchable views.

- **Physics**: This is related to how objects behave in a world, for example, how gravity will affect objects that bounce.

  Generally, virtual objects are built with polygon shapes (known as Mesh) and materials (such as textures and shaders). You can define a **rigidbody**, which is the part of the object that will be subject to motion and physics, and **colliders**, which are bindings for interaction with other objects.

  Lights, shades, and their interaction with the environment and its objects are also important. Audio is similarly important to get right in XR applications.

- **Transform**: This refers to the different changes in the position, rotation, and scale of the objects.

Now that we have defined the components for an XR application, let's see some of the tools that we could use to create one, and let us also look at the testing mechanisms they provide.

## Sumerian [5]

Sumerian is a simple XR development solution by Amazon for the web. It uses **Babylon.js** *[6]* for web rendering and **AWS Amplify** *[7]* for web and mobile development on AWS. Once deployed, it can run on multiple platforms.

Sumerian supports JavaScript but requires little programming, which makes it a good platform in which to get started. However, a bit of knowledge of AWS is desirable to create deployable applications or use AWS capabilities. You need an AWS account to sign in (but you can create a Free Tier one). Besides, since this tool is browser-based, it has fewer requirements from the developer's computer compared to other tools.

One more benefit is that it has a library of assets, including shapes, models, textures, and scripts.

Besides unit testing, it is possible to profile a scene in the browser with Sumerian and Chrome DevTools to check the browser's performance at runtime and measure your app's performance. Additionally, you can emulate different devices to ensure the app UI looks and responds accurately on each of them. For all the rest of the backend testing, you will need to check what testing is available for your specific deployment scenario (there are different options within AWS), and that you have plenty of metrics in place to track the potential issues.

## Unity [8]

Unity is a game development platform that allows you to build 3D and 2D games and deploy them on several platforms. It allows for VR and AR development. Although it is an engine mainly intended for game development, other XR applications can also be implemented using Unity.

Unity supports the C# programming language natively.

One of the benefits of using Unity is that its documentation contains a lot of information, including game tutorials. It is also possible to use the UI and drag and drop functionalities to set up the application, even if your level of coding is not the greatest.

In terms of unit testing, Unity provides a test runner. This can be found under **Window** > **General** > **Test Runner**. We can log object by object or full scenes at runtime. We will learn more about the test runner in the *Tools for XR testing* section.

Besides unit testing on the C# code that the app might contain and some manual testing, we can do some level of performance testing using Unity's profiler window (which should be under **Window** > **Analysis** > **Profiler**). With this tool, you get access to information about CPU and GPU usage, rendering, memory, audio, video physics, lights, textures, and more.

In order to find leaks on the aforementioned values, we could check the UI for some lines in the graphs that do not stop growing and never go back down, or they surpass a value we set up to be our maximum allowed. In the profiler, we can check each component to see what could potentially be causing that particular issue.

We should verify the performance on the specific devices too, as it might be fine in the simulation on the computer and then not work as well on the actual device.

We can choose to show the frame per second on the screen while running and interacting with the application so that we can also verify its general performance while using it.

As with Sumerian, Unity can export to the web. You can check your browsers' performance once you have exported and run your app, as well as device compatibility and emulation.

## Unreal [9]

Unreal is a 3D computer graphics game engine developed by Epic Games. As is the case with Unity, Unreal is not only useful for building games but also for other applications such as simulation, videos, and architecture visualization.

Unreal coding and scripting are done with the C++ programming language. A good number of game companies currently use this language, which makes this engine optimal for them. The reason for this is that the language allows for low-level data manipulation and memory management, making it ideal to work with game consoles and signals between different hardware components. This makes the language extremely useful for communication with devices such as the handsets for VR, although this is not the only way or tool we could use to achieve this.

On the other hand, the C++ language is not as common in the testing sphere or for beginners with automation or coding, and it could be harder to get started with it than with other programming languages. Unreal does provide something called **Blueprints** which is a visual scripting system that uses a node-based interface to allow the user to define objects and interactions. Although it might seem a bit daunting at first, the documentation provides step-by-step guides on how to use it.

Unreal provides capabilities to debug and change variables on the go (under **Window** > **Console Variables**), see real-time data on screen (the `GameplayDebuggingReplicator` class), check the GPU states, collect performance data (from the timing insights window, memory insights, networking insights, and more), and see frontend performance information (with the frontend profiler), among other valuable data about each of our components.

Unit testing could be run without launching the game client (similar to using a headless browser for web applications). This could be useful if we want to automate the execution of those tests from a different program as part of a pipeline, as done previously in the book.

Additionally, it provides a functional testing framework with which we can automate some tests, including stress testing, stats verification, API verification tests, and even configurable screenshot comparison. These are in **Window** > **Test Automation**.

As mentioned before, you should make sure the tests pass on your local machine and within the app simulator as well as on the other supported devices. Unreal allows automation to execute on devices that are connected to your machine or in your network.

In the testing area, Unreal provides a lot of options and stats that might be difficult to learn at first but are very powerful when testing different parts of the application.

Both for developing and testing XR applications, Unreal has tons of useful functionality that allows for very detailed configurations and creations, but it requires a bit more specialization to develop familiarity with. For beginners, Unreal is generally hard to learn (but this might depend on the background of the individual).

## O3DE (Open 3D Engine) [10]

Previously known as Lumberyard, O3DE is an open source 3D game engine (as mentioned earlier, it is not only for games) solution from Amazon. O3DE is cross-platform and capable of creating AAA games (that is, high-budget, high-profile, and high-quality). It allows you to connect games with AWS' cloud capabilities and streaming services.

Functionality can be created with Script Canvas without the need for programming knowledge, or with scripting using Lua (see *[11]* in the *Further reading* section), which is a multiparadigm programming language compatible with C and embeddable in other languages, and C++. Projects can be created with Visual Studio and even through the command line. It has many features including prefabs and physics simulations.

The Lua language is as powerful as C or C++, having the advantage of low-level data manipulation and memory management, making it just as good for device controls and signal handling. It also can be used for functional programming (which could be useful for AI programming *[12]*), procedural programming (which could be useful to program game statuses), and more. On the other hand, it is not very widely known within the testing community, so you will not currently find a lot of detailed or specific information about testing with Lua or many test frameworks designed with it in mind. (This, in turn, makes it extremely interesting for curious, creative minds such as yours, a path waiting to be explored and exposed to the community!)

Among O3DE's tutorials, we can find useful information about unit testing in Ctest (which can have test frameworks such as GoogleTest and PyTest registered into it), a C++ unit testing framework called AzTest, Python test tools for the editor, and other cross-environment testing productivity tools. All this might seem a bit daunting at first, but it presents a lot of possibilities and capabilities that will be useful for your testing needs, including parallel testing and batch test creation.

Additionally, O3DE comes with a profiler tool to monitor network, CPU, and VRAM stats. Other monitoring and debugging options are also available.

O3DE eases the development process with the usage of "Gems," which are packages that contain assets and code that you can reuse in your projects, including AWS, audio, testing, validation, and crash reporting.

The platforms that are currently supported by O3DE are Windows, Android, and Linux.

As with the other platforms, make sure your development, devices, and targeted systems are supported by O3DE.

## Other applications

These applications are not the only ones out there. It is worth mentioning others such as **WebXR** *[13]*, **CryEngine** *[14]*, and **Godot** *[15]* among them. Nor are they the only way you could develop in XR. For example, you could use JavaScript directly using **react-XR** *[16]*. However, it would be best if you find an engine that was originally meant to be used for XR development and provides a good range of tools and capabilities for importing, previewing, and other XR features that make it much friendlier for XR development.

Besides the preceding tools, XR applications generally work a lot with 3D objects. Some tools used to create such objects include **Blender** *[17]*, **Windows 3D Builder** *[18]*, **Adobe Illustrator** *[19]*, **Google SketchUp** *[20]*, and more. We will not be covering them here, as we want to focus on the test side. But keep them in mind if you want to do your own development, and also to ensure you are familiar with the extensions of the files that are produced, as they will be part of your application. However, if you decide to work on XR development, make sure that your XR engine supports your 3D object's format, or that you find a way of converting the objects before you decide to purchase any of the aforementioned tools.

Since the testing examples that we have written in this book are for Unity, let us create an XR Unity application so that we can apply the testing into it later.

# Creating an XR application with Unity

First, you should have Unity installed on your computer. Make sure you are aware of the licenses for each of the versions and that things might be a bit different for your version than for mine. I will be using version **2022.2 alpha**. You will also need Visual Studio installed.

> **Important note**
>
> Please keep in mind that the versions might cause the code to behave differently or might require a little research to make some parts work due to changes in the API. For example, our reviewer had issues with the `onMouseDown()` method, which were resolved by updating their installed version to 2021.3.16f. The purpose of this book is to showcase how testing in XR works. For more information on the particulars of Unity or any other tools mentioned, please check our references at the end of the chapter.

Once you have all your requirements ready, let us create an XR application:

1. To create a new project with Unity, select **New project** from the initial screen or click on **File > New Project**. For older versions of Unity, you could select **3D projects**, while for the new version, you can install the AR or VR types, as well as Mobile 3D. If you installed **Unity Hub**, you can also create an app from there, by clicking on **New** and selecting your version of Unity.

2. Let us make something simple for our first 3D app: a time machine. We will only use a rectangular cube as our time machine, and a smaller one for its door. Feel free to play around, adding audio, backgrounds, textures, lights, and more.

   To add a new 3D cube, right-click within the **Hierarchy** area, click on **3D Object**, and then click on **Cube**. This might be different for your version of Unity, so make sure you check the documentation for your version:

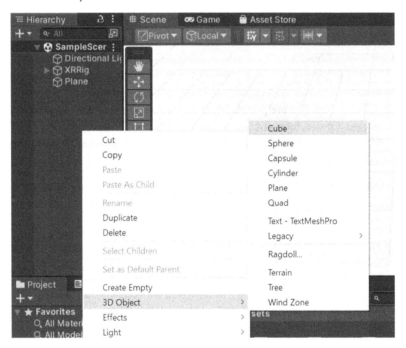

Figure 10.2: Menu Hierarchy > 3D Object > Cube

3. Click on the menu that appears in the scene to change the object's size and position. You can move through the scene with the hand icon of the same menu, and in order to rotate yourself throughout, press *Alt* at the same time as you click. Repeat the process to create the second object, or you can right-click on the object in the **Hierarchy** area and duplicate it so that the new object is already in place.

4. As the idea is to have a door that opens, for better results, you can have four 3D boxes, so that three of them are walls, and then add two more on the top and the bottom to signify the ceiling and floor. If you want to color them, you can create a new material (right-clicking in the **Asset** area should show you that menu) and change the albedo of such material. Then, drag it and drop it on top of the asset in the scene to add it to it (although you can also do that from the menu).

We put a different color for the door and also added a small capsule on top of the box that we could use as a light further on if we wish to:

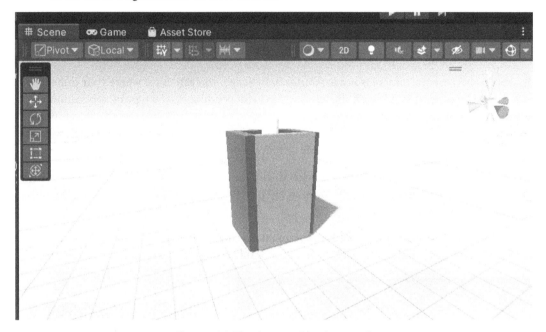

Figure 10.3: The time machine box result

There are plenty of things we can add to this application; we just want to make it simple as our starting point. We will add some functionality in the *Unit testing in Unity* section.

Keep in mind that in Unity, we can create and change all the objects by code too. We have now created our time machine, feel free of adding your own design, adding backgrounds and building on top of this and, if someone tells you that's a small box for a time machine, feel free of telling them that it is bigger on the inside

> **Note**
> You can add an empty object to wrap all the components of the time machine, so you can move it around as a block.

Now that we have our application, it is important that we configure the camera so it will show what the user sees

## The camera

If you want to test the application, even though it does not really have any functionality, notice the **Game** tab, which is next to **Scene**. In this tab, we can change **Game** for a simulator so that we can see how the app runs on different devices. However, the chances are that when you go to this tab, you won't be able to see our time machine.

This happens because the camera is not *looking at it*. In **Scene**, you can move where your camera is, which should be defined as **XRig** on the left-hand side (depending on the type of application you selected to create, it might also be **Camera**). If you still cannot find it, try to expand the objects under **Hierarchy**. Then, feel free to move it and rotate it until it looks at the time machine, which we will then be able to see when we move to the **Game** or **Simulator** tab.

This simulator does not have default XR behavior, that is, the camera is not going to move unless you add some script for it to do so. That's OK for now, but just make sure the camera is pointing roughly at the time machine.

You might also want to add one or more XR plug-in management engines to support the simulations under **Project Settings** (which is accessible from **File** > **Build Settings**) as per the following screenshots:

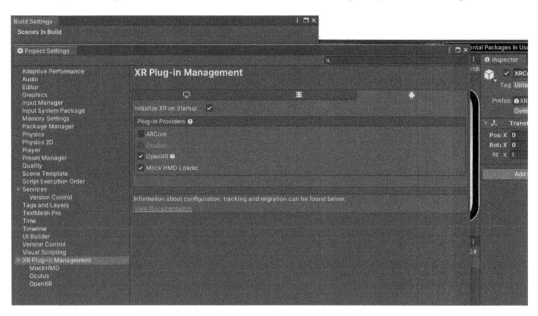

Figure 10.4: The Build Settings menu -> Project Settings -> XR plug-in management

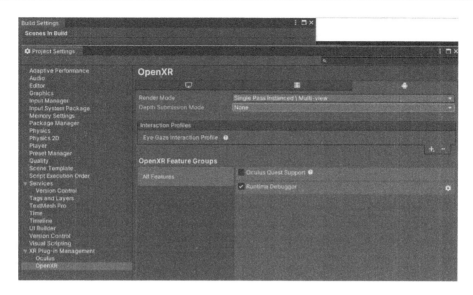

Figure 10.5: The Project Settings menu -> OpenXR

Finally, if we want to interact with our application and be able to execute it on our actual device, we should set up our platform under **File** > **Build Settings**. We can change to the platform we want by clicking on **Switch Platform** after selecting the one we would like to switch to. Make sure you have the package name selected in **Project Settings**, as per the following screenshot. After this, you can click on **Build And Run** to install it on your connected device:

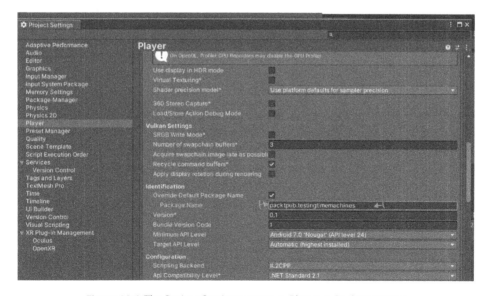

Figure 10.6: The Project Settings menu -> Player -> Package Name

> **Note**
> Make sure your device has been set to developer mode and that it has allowed the USB to connect to your computer prior to debugging on it.

In the next section, we will see a couple of tools that could be of use for testing XR applications, although others might arise by the time you read this book. Therefore, as usual, feel free to do your own research and try to find the best tool for your specific application.

# Tools for XR testing

As we saw in the previous section, while developing an XR application, there are multiple potential roads that you could take. Each of the platforms has its own internal ways of doing unit testing, performance checks, and other validations.

Additionally, we have seen that there are many types of testing we should take into account, some of them very specific to XR. The most challenging part of testing an XR application is performing some of these tests, as they require different test users and a lot of manual effort.

Lucky for us, there are some automation options that facilitate these tests, although they are platform-dependent. That said, this is a fairly unexplored field and more discoveries would be helpful. Maybe you can dare to create some tools in this area?

In this section, first, we are going to see how to create a simple unit test using Unity's test runner. Then, we will explore some tools for XR automation. The examples will be done in Unity, as it is where I currently have more expertise, and several books could be created to cover every possibility on every one of the engines.

First of all, let us see how unit testing works for XR.

## Unit testing in Unity

The Unity Test Framework package (formerly known as the Unity test runner) executes your test code against standalone, Android, or iOS platforms. It supports play mode or edit mode, so make sure you check the documentation to get the best of both modes and that you are running your actual scripts, so you don't cause a false positive. It is always good practice to run your tests in both modes.

The test framework that it uses is NUnit.

Let us write some code for the application that we created in the last section and add some unit testing to it. The goal is to open the door of the time machine.

We create a new script in the **Assets** area and type in the following code:

## Opendoor.cs

```
using UnityEngine;
public class Opendoor : MonoBehaviour {
    public GameObject door;
    public bool doorGoingUp;
    void OnMouseDown(){
        doorGoingUp = true;
    }
    // Start is called before the first frame update
    void Start() {

    }
    // Update is called once per frame
    void Update(){
        float y = door.gameObject.transform.position.y;
        float x = door.gameObject.transform.position.x;
        float z = door.gameObject.transform.position.z;
        if (doorGoingUp &&
            door.gameObject.transform.position.y <= 6){
            y += 0.05f;
        door.gameObject.transform.position =
            new Vector3(x, y, z);
    }
}
```

This code has a method to set a variable to true when the door is clicked on (the OnMouseDown method). Then, the update method that runs every second for our application will update the door to go up as long as that variable is clicked on and the door's vertical position does not surpass 6. You might need to adjust this number for your application depending on the position of the machine. In the GitHub code, we have added a variable to set it up from the editor.

Note that we have created a public variable, called door, that obtains the door object. We can attach the script to any game object (for example, to the door or to an empty one) in the hierarchy area. We can do that from the inspector of the object by clicking on **Add Component** > **Script** and typing Opendoor into the search. Then, we need to drag and drop the door object from the hierarchy area on the variable that we created, as follows:

Figure 10.7: The Door object added to the Door variable in the Opendoor component script

Alternatively, we can create a new tag for the door object and assign the tag to the door object from the inspector menu that appears on the right-hand side after clicking on the door object in the hierarchy menu (the hierarchy menu is on the left-hand side):

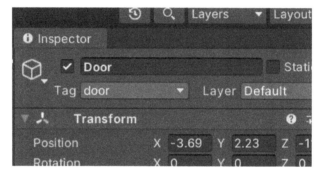

Figure 10.8: Adding a tag to the door game object

Then, we would not need the `GameObject` variable, and we can use the tag to find the object within the update method, such as in the following line of code:

## Alternative line to get the door object:

```
GameObject door = GameObject.FindGameObjectWithTag("door");
```

Let us continue with this second method as not having this variable will help us later on when we look at automation in Unium (in the next section).

> **Advanced note**
>
> Using `onMouseDown` could be useful for PC games. However, when it comes to XR, we might need to interact with a handset or have a device without one. In the latter case, we need to make sure the camera is centered on the object. You can find more information about all of this in Unity's documentation. Check the `Raycast` and `RaycastHit` methods to find out more.

Now, we can test the application by going to the **Simulator** tab (or **Game** if you haven't switched it to simulator) and clicking on the play button. By clicking on the door, you should see it rising. Play with the variables to make it rise slower.

In an ideal world, before writing the actual code, we should have added some unit testing. Let us add this now. First, let us add a test folder by clicking on **Window** > **General** > **Test Runner** > **PlayMode** -> **Create Test Assembly Folder**.

Then, we create a test script by clicking on **Create Test Script inCurrent folder**. We can then edit and rename this script to match the following pieces of code (`DoorTestScript.cs`).

First, we import the libraries we need:

## DoorTestScript.cs (Importing libraries)

```
using System.Collections;
using Nunit.Framework;
using UnityEngine;
using UnityEngine.TestTools;
using UnityEngine.SceneManagement;
```

Then, we create the class and add a test. In this case, we are adding a test of the `IEnumerator` type because we are yielding the result at the end. This helps us wait for the frames or domains to be loaded.

In the test, first, we wait for the scene to load (since this is a play type of test), so all the objects are present and we can interact with them. Next, we find the object we want to use for the test (the door) and retrieve its position. Next, we perform a click on the object by sending it an `onMouseDown` message. Finally, we wait for the domain to be reloaded and check that the position has effectively changed.

Let us see the code for that. Note that we renamed our original scene to **TTM** (short for **testing time machines**):

## DoorTestScript.cs (Class after libraries import)

```
public class DoorTestScript{
    [UnityTest]
    public IEnumerator DoorTestScriptWithEnumeratorPasses() {
        SceneManager.LoadScene("TTM", LoadSceneMode.Single);
        yield return new WaitForDomainReload();
        GameObject door = GameObject.Find("Door");
        Vector3 position = door.transform.position;
        Debug.Log("Initial position " +
                door.transform.position);
```

```
            door.SendMessage("OnMouseDown");
            yield return new WaitForDomainReload();
            Assert.AreNotEqual(position, door.transform.position);
            Debug.Log("Final position " + door.transform.position);
            yield return null;
        }
    }
```

For this to work, we need to set up the scene to load in the **Build Settings** window (**File** > **Build Settings** > **Add Open Scenes**):

Figure 10.9: Adding the scene to be loaded

This is not the only way we could have achieved these results and performed the tests; we considered this to be an easy way to explain it to beginners. Feel free to play around with Unity to find other solutions that adjust better to your needs.

Now that we have learned how to create unit tests, let us see how we can automate the actions to avoid repetitive manual testing.

## XR automation

Understanding how unit testing works on XR helps us to understand where we could get started with automated testing in XR. As mentioned earlier, automation is platform-dependent. Not only does it depend on what engine you are using for development, but it also on what device the application will run on.

For XR, there are two types of automation: hardware and software.

In hardware automation, we check the signals that the different devices send to our program, and we simulate such signals without the need for a device. Alternatively, we can copy what we get from our current device for recording and playback functionality. Generally speaking, frameworks for hardware automation or simulators are harder to implement, as they require a deep knowledge of signals and a programming language that allows this sort of communication, such as C.

In software automation, we move across our application as though it was done by the user, interacting with the object through their interface, DOM system, or with the use of the camera or player. This can be achieved with a high-level language and more general knowledge.

In XR, there are different elements other than visuals that could also be hard to automate, not only from the hardware's point of view but also from the software's point of view. For example, verifying things such as audio cues, motion sickness, or accuracy of the objects and lighting are very challenging to automate. We have highlighted this several times throughout this book, but let us do it one more time.

**Automate what makes sense to automate!**

Automate whatever will save you time in the long term by being automated (considering the time spent in creating the automation). Avoid spending too long automating things that only need to be checked once or in cases where the automation will not be reliable in short time.

For example, if your company works with a specific piece of XR hardware and it makes sense to spend resources on it, then automate it (by automating its signals exchanges with the other devices). Otherwise, try to sort it out with software automation or invest in tooling to automate your needs.

Let us see two ways in which we can automate our tests using software. We will use Unity and the project we created earlier for these interactions.

## Unium [21]

Unium is a library for Unity that lets you manipulate the scene for automation purposes. It works over an API with a web server that is embedded into the game. This is kind of similar to how Appium or a remote web server would work. The default port that Unium uses is *8342*.

You can find it on GitHub at `https://github.com/gwaredd/unium` or on the Unity asset store (**Window** > **Asset Store**). If you use the Unity's asset store, you will need an ID and password for downloading content. All the information about the library is in the previous GitHub link. For newer versions of Unity, you can add the package directly from the GitHub URL by clicking on **Window** > **Package Manager**. Then, click on the + symbol to access that option. Refer to the GitHub URL for more information on alternative ways of installing Unium.

Once installed, we should have a `Unium` folder on the **Asset** side of our project. In order to use Unium, we should add the script to an object in the scene. For example, we can use an empty object. Then, we add the `UniumComponent` script to it (by adding a component or dragging the script files to the object).

> **Note**
>
> We could also add the `UniumSimulate` script for some extra functionality, such as clicking and dragging.

After this, we can open a browser and perform calls directly on it. For example, if we want to move the camera (which is crucial for XR applications and VR in particular) to position 1,1,0, we could do something like this:

```
http://localhost:8342/q/scene/XRRig.Transform.position
={'x':1,'y':1,'z':0}
```

This rotates XRRig. If you have another object or just your main camera, you should use that name instead. Try using the following to see your available objects:

```
http://localhost:8342/q/scene
```

This will inform you of the objects of the scene, so it is easier to know what to call. If you are still unsure about what objects to interact with, you can even use wildcards or queries. Refer to Unium's documentation for more information about this.

If we want to rotate the camera, we can use `eulerAngles`:

```
http://localhost:8342/q/scene/XRRig.Transform.eulerAngles=
{'x':0,'y':180,'z':0}
```

This rotates the camera over the $y$ axis by 180 degrees. As of the time of this writing, the only rotation you can perform with Unium is with Euler angles (as opposed to quaternions). Be careful not to land in the gimbal lock, as we explained in a previous chapter.

We could rotate the camera more slowly by setting the rotation speed of the object, but you would need a bit of extra coding for this.

If you use a handset, you should be able to perform actions on it in order to interact with your applications. Otherwise, it is recommended that the application has some areas into which the player can move. These are called waypoints. If we had some of them in our application, we could also interact with them in Unium to make our way forward.

As another example, we can use Unium to perform the click on the door, by sending the same message that we sent with the unit tests, like this:

```
http://localhost:8342/q/scene/Door.SendMessage(OnMouseDown)
```

All these browser calls are nice to have, but they can get very manual, which runs counter to automation. For that reason, Unium allows the use of several scripting languages (for example, C#) to automate the calls to perform one after the other. For example, if our camera was not in position, we could turn it around and then click on the door to verify its position.

Here is an example of how to do this:

## UniumTests.cs

```csharp
[TestFixture]
public class Unium_Functional {
    [Test]
    public async Task CheckDoorBehaviour() {
        using( var u = new WebsocketHelper() ) {
        // connect to game
        await u.Connect( TestConfig.WS );
        dynamic camera = await u.Get("/q/scene/XRRig");
        await u.Get("/q/scene/XRRig.transform.position={\"
                   x\":1, \"y\":1, \"z\":1}");
        await u.Get("/q/scene/XRRig.transform.position={\"
                   x\":180, \"y\":1, \"z\":1}");
        dynamic doorPosition =
            await u.Get("/q/Scene/Door.transform.position");
        await u.Get("/q/Scene/Door.SendMessage(OnMouseDown)");
        dynamic doorPosition2 =
            await u.Get("/q/Scene/Door.transform.position");
        Assert.AreNotEqual(doorPosition, doorPosition2);}}}
```

The preceding code can run in parallel with our application's code on its own project. Make sure you are playing the application or simulation before executing the tests. We could use other languages, such as JavaScript or Python, and diverse testing frameworks besides NUnit.

Since our application is small, we have just barely increased the complexity of our unit test for our functional testing, but keep in mind that the idea is to use this to help us reproduce more intensive user scenarios, flows, and repetitive tasks.

Now, let us see how to do something similar with a different tool for comparison purposes.

## Airtest Poco [22]

Airtest Poco is an automation framework developed by the company Netease. It contains three main parts:

- **Airtest**: This is the UI automation framework based on image identification
- **Poco**: This is the UI automation framework based on UI control recognition
- **AirtestIDE**: This is a cross-platform testing editor

While we saw an example of this tool in *Chapter 5*, *Testing Automation Patterns*, we wanted to highlight its capabilities for XR testing, too.

Let us create a VR automation using the Airtest project:

1. To use Airtest with Unity, we need to download the Unity3D folder from `https://github.com/AirtestProject/Poco-SDK/tree/master/Unity3D` and add it to the assets in our Unity project.

2. Remove the `fairygui`, `ngui`, and `uguiWithTMPro` folders, so there are no conflicts about which scripts to use.

3. For this automation, we need the camera to be contained within two objects, so if you have XRRig, it should be fine; otherwise, make sure you create two empty objects, one as child of the other one, and have the camera as a child of the inner child, as shown in this picture:

Figure 10.10: Camera structure

4. Now, we can use `AirtestIDe` (available at `airtest.netease.com`) to add Python code to automate the movements that we want to do. For example, if we want to rotate the camera, we could use the following:

```
vr.rotateObject(-10, 0, 0, 'XRig', 'Camera Offset', 0.5)
```

5.  Now, if we want the camera to look at our time machine, we could use the following:

```
vr.objectLookAt('Door', 'XRig', 'Camera Offset', 5)
```

We can select the speed of movement with the latest input in the call (which is 5 in the previous line).

This type of automation could be very handy if we need repetitive actions or if we want to automate a complete game scenario.

6.  To rotate the camera, go to the test machine, click on the door, and verify that it went up. The code should look as follows:

## DoorAuto.py

```python
from airtest.core.api import *
auto_setup(__file__)
import time
from poco.drivers.unity3d import UnityPoco
from poco.drivers.unity3d.device import UnityEditorWindow
dev=UnityEditorWindow()
addr = ('', 5001)
poco = UnityPoco(addr, device=dev)
#note: use device=UnityEditorWindow() to test in unity
# dev = connect_device('Android:///<serialno>') to
# test in Android
vr = poco.vr
poco = poco()
position = poco('Door').attr('position')
vr.rotateObject(-10, 0, 0, 'XRig', 'Camera Offset', 0.5)
vr.objectLookAt('Door', 'XRig', 'Camera Offset', 5)
count = 0
while(!vr.checkIfUnityFinished() && count < 10):
    time.sleep(2)
    count = count + 1
poco.click()
assert poco('Door').attr('position') != position
```

Now that we have seen a couple of examples of software XR automation, let us review the differences with hardware automation and when each would be more interesting or suitable for our application's testing.

## Hardware automation

While hardware automation is closer to the final way the user would interact with our application, it is possible that we will still need to configure some software parts in order to get full end-to-end automation. Another difficulty here is that the signals might vary from device to device, so it is possible that we need a different set of calls and emulations for each type of device we have. Also, this means that we need powerful systems to even try this out, unless we can find some cloud providers for it (as discussed in the previous chapter). For the purposes of this book, and this chapter, we will not go further into hardware automation, but feel free to do some research in case you need to use it.

## XR emulators

For XR hardware signals testing, it would be useful to have an emulator that would let us use our keyboard or mouse as if they were the XR device. This is useful for teams with several members that want to test their changes before using the actual devices, so we can have one device for the entire team rather than one per member.

However, when it comes to automation, the emulator is not the only part needed. We also need to have the capabilities to record and play back the actions that we have taken or to program such signals to be sent in a specific order.

An added difficulty for this type of automation is that it requires different code/signals for different supported devices, which could complicate the automation code, but results in higher-quality automated tests.

Keep in mind that this emulator must work on our development platform; otherwise, we would have to deploy it to whichever sharing platform we have in order to make use of it. Some sharing platforms might allow you to deploy the app or game on their store without displaying it to the public, but most of the time, deploying there would mean that the app will also be shared with your users, which would be quite late for testing purposes.

The key questions to ask to understand whether we need to use a hardware emulator to test our app are listed here:

1. Does your app generally and intensely on hardware or does it have specific features that can be tested alone through the usage of hardware? If so, you should probably invest in an emulator.

2. Are there specific repeatable actions that need to be checked or are there things that constantly change and are difficult to repeat? If so, you need a way of recording and playback (being emulator or software).

3. Could software automation cover the most repetitive actions? Otherwise, we should invest in an emulator.

4. Do you work specifically with one device, a few hardware devices, or with many different hardware devices? Might need to invest in more than one emulator, if so.

5. Are there already hardware automation tools that would work with your development platform and with your main supported devices? It would probably be better to use those tools than reinventing the wheel, if we can afford it.

6. Are you planning on creating several versions and screens with similar actions or is this a one-time-only deployment? We could repeat automation if the actions are repeatable.

Due to the complexity and specificity of hardware emulation, we will not cover it in this book, but we still wanted to give you the guidelines in case you wanted to look deeper into it.

Now that we have reviewed the entire spectrum of XR, let us review a concept related to XR that has surfaced recently so that we can see it from a testing perspective.

# The metaverse

The metaverse might sound like something out of a science fiction movie, but it is just another way of connecting and interacting with each other over the internet, in an immersive virtual world thanks to the use of XR devices. Therefore, from a testing perspective, testing of the XR devices, actions, and test types is needed when it comes to sites or applications that allow for such interactions and usage.

We have mentioned some concerns of testing when it comes to social interactions, such as personal space and security. We should work extensively on these concepts when it comes to the metaverse, while kinetic movement and motion sickness will still be important.

Furthermore, some applications for the metaverse require not only the input of how the user is moving physically but also how it looks. For example, there are applications that allow the user to try on clothes virtually, measure objects, or simulate placing them in their homes. These sorts of applications require analyzing camera inputs (such as videos or photos) that could be very private and, as such, would require stronger measures to ensure that data privacy is maintained.

I will also argue that we should provide users with reminders to take regular breaks from the XR experience and make sure that external alarms are extended to virtual worlds. Unfortunately, this contrasts with the goal of many companies to create immersive experiences in order to hold the user's attention for as long as possible.

Most of the testing mentioned in this section involve things that cannot be easily automated and might require careful planning, along with manual testing by different people, besides a good platform for user support so that users can report any issues that we might not have encountered during testing.

In conclusion, while testing in XR has a series of considerations that we have reviewed throughout this chapter, testing in the metaverse adds important social and ethical considerations that are crucial for your application to work well and to protect your users.

# Summary

In this chapter, we started by reviewing the different concepts and related applications of XR, including VR and AR, and what particularities they have in terms of ensuring quality. Then, we navigated through the testing pyramid for XR applications and found common points with normal applications, as well as differences and unique aspects with XR ones. One particularity of these applications is the issue of gimbal lock and the use of quaternions to best avoid it.

We continued the chapter by going over some tools for XR development and testing in XR. These tools are not the only ones available, but we consider them as a good starting point for those who want to start with XR or those who want to understand the concept of testing with XR.

Then, we briefly went over the concept of the metaverse and how to apply everything learned in this chapter to it.

After reading this chapter, you should be able to grasp the XR concepts needed to create and test your own XR tools and applications.

In the next chapter, we will go over some applications that are difficult to test, and some real-world examples in which testing could be challenging, and we will see what we could do to ensure their quality.

# Further reading

- [1] If you are interested in learning about Pokémon go, you can find more information here: `https://www.pokemon.com/us/app/pokemon-go/`.

- [2] See this video explaining the gimbal lock effect: `https://www.youtube.com/watch?v=zc8b2Jo7mno`.

- [3] See `https://apollo11space.com/apollo-and-gimbal-lock/` and `https://www.hq.nasa.gov/alsj/e-1344.htm` for more information about apollo considerations with gimbal lock.

- [4] Find the source code of Apollo 11 here: `https://github.com/chrislgarry/Apollo-11/blob/master/Luminary099/IMU_PERFORMANCE_TEST_2.agc`.

- [5] You can find more information about Sumerian on its website, including tutorials and other documentation: `https://aws.amazon.com/sumerian/`.

- [6] Take a look at Babylon.js (`https://www.babylonjs.com/`) for web rendering.

- [7] Take a look at AWS Amplify (`https://aws.amazon.com/amplify/`) for web and mobile development on AWS.

- [8] Find out more about Unity here: `https://unity.com/`.

- [9] Unreal's website also contains documentation and tutorials at `https://www.unrealengine.com/en-US/learn`.

- [10] For more information, such as tutorials, documentation, video explanations, and a link to get started, visit `https://www.o3de.org/`.

- [11] For more information on using Lua, visit this site: `www.lua.org`. Check `https://www.packtpub.com/product/lua-quick-start-guide/9781789343229` for a quick start guide and `https://www.packtpub.com/product/lua-game-development-cookbook/9781849515504` if you are interested in how to use Lua for game development. You can also check out this book: `https://www.packtpub.com/product/lua-quick-start-guide/9781789343229`.

- [12] See `https://hub.packtpub.com/what-makes-functional-programming-a-viable-choice-for-artificial-intelligence-projects/` for more information about functional programming and why it is frequently used for artificial intelligence.

- [13] For more information on WebXR, visit `https://immersiveweb.dev/`.

- [14] For more information on CryEngine, visit `https://www.cryengine.com/`.

- [15] For more information on Godot, visit `https://godotengine.org/`.

- [16] For more information on ReactXR, visit `https://github.com/pmndrs/react-xr`.

- [17] Information about Blender can be found here: `https://www.blender.org/`.

- [18] More information about Windows 3D builder can be found here: `https://support.microsoft.com/en-us/windows/how-to-use-3d-builder-for-windows-e7acb10c-d468-af62-dc1d-26eccd94fae3`.

- [19] The Adobe Illustrator documentation can be found here: `https://www.adobe.com/products/illustrator/free-trial-download.html`.

- [20] For more information about Google SketchUp, visit `https://www.sketchup.com/products/sketchup-for-web`.

- [21] The Unium documentation can be found here: `https://github.com/gwaredd/unium`.

- [22] For more about the Airtest project, go to `http://airtest.netease.com/`.

# 11
# How to Test a Time Machine (and Other Hard-to-Test Applications)

Throughout all the previous chapters, we covered different technologies, the knowledge of which can help while testing difficult apps. These concepts should cover most of our quality needs. However, sometimes it happens that we face applications and systems out of the norm, and guides as such the ones provided in the previous chapters would be insufficient for our needs.

In this chapter, we will focus on challenges that are particularly difficult to solve and provide solutions that you might find useful for your own challenges and systems.

Before we dive deep into the different solutions to ensure quality on difficult applications, let us highlight the importance of understanding what's worth spending time and resources to automate and what's not. If the effort spent on automation or even the solution of a problem surpasses, in the long run, the effort of a different workaround, such as performing a manual validation, we should not continue with that automation. This might seem logical to most people, but it is not easy to remove the opportunity of figuring out an approach that feeds both the mind and our curiosity. While researching and thinking outside the box is good sometimes, other times it can be better to simply reuse an existing solution, so keep that in mind, and avoid falling to the temptation of curiosity to the detriment of finding a solution.

Rather than reinventing the wheel, there are a number of apps available that can be helpful, and you could help in making some of them by contributing to the open source space, which needs creative minds.

That said, the goal of this chapter is to inspire you to think outside of the box, and it provides help and tips to test challenging applications. On top of this, this chapter will also help you build on concepts we discussed in previous chapters and consider an overall architecture with everything we have reviewed in this book so far.

In this chapter, we are going to cover the following main topics:

- How to test a time machine
- Challenges in game testing
- Automating the *impossible*
- Automating for a higher purpose
- Last-minute tips and best practices for your automation
- Architecting a system

## Technical requirements

The first section of this chapter is merely hypothetical, included so that you understand the different approaches to testing before starting with any programming or deciding on the architecture of a system, so there are no technical requirements for it. For the rest of the chapter, some degree of programming skills is recommended to get the best out of the examples.

Whilst this chapter is written with a QA/SDET role in mind, as applications shift left, developers may also find this chapter interesting. Furthermore, if you are trying to get developers more involved in testing, this is one of the chapters you would ideally show to them, as it, hopefully, will trigger their building instincts and curiosity so that they also contribute to the creation of amazing applications to improve the quality of systems.

This chapter uses examples written in various programming languages. We recommend reviewing and working with different languages as a self-growth exercise; we provide an implementation for other languages in our GitHub: `https://github.com/PacktPublishing/How-to-Test-a-Time-Machine/tree/main/Chapter11`.

## How to test a time machine

You are probably wondering how we came up with the unique title of this book as well as this chapter. It all began with an idea I wrote on my blog – if a time machine were to exist, how would you test it?

This comes from a typical test interview question in which the interviewer would ask the interviewee how to test some object, and try to identify/organize the cases in different types of tests. The question might appear silly, but it actually tells a lot about the interviewee's testing capabilities, out-of-the-box thinking, curiosity, inventive and testing knowledge.

Therefore, feel free to take a pause from reading, take a notepad, and do this exercise – *if time machines were to exist, how would you design the testing for them?* Start writing everything that comes to mind. Then, keep reading to see whether there was something that we or you have missed.

At the end of the day, when it comes to testing an application, you may as well be asking yourself the same question. This can help you prioritize tests based on the available resources.

There are many different approaches to this sort of question, but we could categorize them into one of these three:

- **Bottom-up**: We try listing the different test types and try to find examples in our application that would fit each one of them. This usually works quite well in interviews, as it helps us give a more structured answer.

- **Top-down**: We start thinking about ways we could test the application in general or its basic functionality, and then go into more detail and add other tests. We categorize the tests by types as we go along. This is usually what we end up doing when we are pressed for time on the delivery of an application.

- **By analogy**: We could compare an application with some other application that we have seen in the past and think of how we or someone else tested that one. We could also mix bottom-up or top-down methodologies within the analogy.

## By analogy

Let us think of the testing of the time machine from analogy – I imagine that traveling across time could be similar to traveling across space; therefore, we could compare a time machine with a spaceship or a rocket. In my blog, I started with this example and I checked the actual code of Apollo 11, which happens to be available on GitHub *[1]*, to see what sort of testing NASA did for it. Then, we can add tests to our plan by comparison. Of course, this code was written in the *Assembly* programming language, and nowadays, it is likely that rockets use higher-level languages for their programming, but, still, it's good to have an example to get started.

In this case, NASA found it important to check for arithmetic errors, signals, and even gimbal lock checking (we learned about this concept in *Chapter 10, Travelling Across Realities*).

These sorts of low-level and space-machinery issues would have not been so obvious if we just went with the two first approaches, and we might have found ourselves having to tackle issues as they arose. This is never a good position to be in. When developing an application, the ideal scenario is that we think of everything that could go wrong ahead of time so that we can develop with those issues in mind beforehand. If you are facing a hard-to-test application, do not forget to do some research in the field and about other similar applications, and see what solutions were implemented for those.

You might think your app is unique, but it will surely share some aspect or another with another available application, as most inventions come to fruition as a combination of various other existing things.

Now that we have reviewed how to think of an application by analogy, let us continue with my favorite approach – bottom-up.

## Bottom-up

For this approach, instead of looking at another application for inspiration, we can turn to the tests and secret passages in the pyramid. For this, we could use the pyramidal cheat sheet that we featured at the end of *Chapter 4, The Secret Passages of the Test Pyramid – The Top of the Pyramid*. Then, we will see whether we can cover every aspect of the application with it (spoilers – we won't!) In this section, we will see some other type of testing that was not previously mentioned in the book.

### The base level of the pyramid

The tests in the low level of the pyramid are the hardest to plan when looking at a general problem. The reason for this is that their nature is rather specific. We can plan to have unit tests, code coverage, code reviews, optimize the code, and test databases. However, in order to have specific tests, we should be planning to write specific code. Time machines have not yet been invented (at least to our knowledge). Therefore, we would have to plan specific features in order to devise these tests.

That said, since we used the analogy approach, we could keep going along those lines and specify tests for features that we know exist in rockets and the like. Keep in mind as well that a time machine is not only a software product but also a hardware one, and therefore, we would have to make several decisions on the physical design, which would require its own set of tests.

There should be a special mention of two things here:

- **Mocking**: Testing time travel would be very expensive, so mocking (covered in *Chapter 2, The Secret Passages of the Test Pyramid – The Base of the Pyramid*), stubbing, and so on are especially important to get right in our system.
- **Database testing**: Databases require stability and persistence. Both of these concepts are related to order and time. If we have databases inside of our time machine, we could run into problems because of the dates and times at which the data is recorded and read, so we would require a good test plan for them. For example, if we enter some data and then travel to the past and check the past 3 days of data, we will be missing that data. The time in our database should be relative to some clock on the machine, but we should still be capable of retrieving the current date that we have traveled to.

### The middle level of the pyramid

Let us think of examples of tests for each of the components at the middle level of the pyramid.

#### Integration tests

Let us try to test the integration of our machine with the environment and other objects, even though it would not be a typical integration test as we have seen previously, since this is a special system. Let us categorize these tests as integration, as follows:

- Place a physical object, such as a chair, where the machine is positioned before sending it for a trip. Move the object away, then move the machine to that same spot, and test traveling to a time before we positioned the object. Verify that neither the object nor the machine break.

- Test traveling to a time when the object is still in place; verify that neither the object nor the machine break. Verify that there are no other alterations in space-time.

- Travel twice to the same exact time and place. Verify that the time machine maintains integrity.

- For the next check, we would have to know whether the time machine can travel between different universes. Two of the most discussed theories about how time traveling can be achieved are as follows:

  - By using negative speed, in which case we won't find ourselves in different universes.

  - By using wormholes, which could mean traveling across universes. Therefore, we have the need for such a test and other specifics about universes and guaranteeing we are in the right one.

- If we have something inside the machine, does the object conserve its integrity? Can it safely travel with the machine? Does the machine also conserve its integrity? Are there any other space-time issues?

### Service tests, API tests, and contract testing

The following tests are some other examples we could think of for the middle point of the test pyramid:

- We should test whether there is a way of communicating across time to verify whether we can communicate efficiently and as expected across different times and spaces.

- We should make sure that all the pieces of the machine communicate effectively. Both hardware and software should be considered.

- Try to communicate back and forth in time between the pieces. We might have issues if we try to communicate some message with an older or newer date than expected.

- Daylight-saving or specific dates (such as February 29th) changes could also affect any communication that requires us to validate time.

### System testing

If our machine is composed of different pieces, we should make sure they are connected well and that we test each system separately and together:

- Check that all the systems work well together.

- Check that all the pieces are aligned when traveling across time (maybe some material needs a different process in order to do so) and work well individually as well.

## The top level of the test pyramid

Now, let us focus on the top level of the test pyramid and try to come up with tests that would be useful for us at this level.

### End-to-end testing

Let us first list some end-to-end tests:

- **Happy path**: Go forward in time and then backward in time and measure the success, verify no changes occurred, there were no breaks in time-space, and so on.

- **Verify the safety of the machine**: Introduce an object before a person tests the machine. It should be an organic object that has a known life span so that we can check how long time it has been inside the machine. As the machine goes forward or backward, the state of the object should stay constant while the surrounding world changes.

- **Test time boundaries scenarios**: Go back to the start of the earth (or universe) and forward to its end. We might need to include some exploratory tests here. What happens if we travel just before or just after? Does time always exist? Can we travel to a point there *is no time*?

  - What's the minimum time that we could travel? And the maximum?

  - How specific can the machine be about time? Days? Seconds? Nanoseconds?

- **Test space boundaries**: if the machine also travels in space.

- **Others**: Test any other user interface-related feature, including lower-priority tests.

We can also consider/categorize some of the previous tests as usability testing. We generalized them here as end-to-end for simplification.

### User interface

In these tests, we should think about things related to the user interface, such as the following:

- Test the minimum and maximum size of the machine

- Test each of the controls in the machine's control panel function appropriately

### Accessibility

Let us list some tests related to accessibility:

- Check that it is possible for everybody to travel in the machine. If not, what should the requirements be?

- Check the control panel of the machine is easy for everybody to manage.

- Check that the seats are comfortable for this type of travel and suitable for everyone.

## Performance

Let us list some tests related to the performance of the time machine:

- Check how many times the machine can go backward and forward in time before breaking or destabilizing
- Take some of the previous tests (for example, the integration ones) and repeat them several times to validate the performance of the time machine
- How many people/objects can the time machine handle?
- For each component, measure their performances going back and forth in time and try to find leaks and issues
- Does the time machine require a cooldown period before traveling once again?

## *Other testing*

As previously mentioned, the pyramid represented in this book, although quite extensive, is not enough for all the potential tests of the time machine. There are a couple of extra type of tests that we could add here.

## Exploratory testing

Since humans have never (to our knowledge) traveled in time before, there will be a lot of things we do not know about it. We have discussed this sort of testing in the aforementioned end-to-end discussion about boundaries. Some hard-to-test applications, such as this one, have very ambiguous features, and we are not sure how they are meant to behave. In these cases, and when we cannot ask anybody in our team or company about similar requirements, the best thing to do is to explore its behavior. On occasions like this, when the product is already finished and we want to get a better understanding of it, while trying to find potential bugs, exploratory testing would be a good approach:

- In order to test whether something will change the future or even produce a parallel universe, we could try changing just a very small thing in the nearby past – something that could be easily undone
- We should think of other ways of testing possible paradoxes, such as a time traveler meeting their past or future self

**Security testing**

Let us consider some tests that could align with security testing:

- Make sure that the time machine does not harm the traveler. For example, insert devices to measure whether the cabin is habitable before sending people. Test first with robots, and then with some fruit or a decaying object, and so on.

- If anything goes wrong, can the travel be interrupted/reverted? What would be the risk for the passenger in this scenario?

- Test that there is a safe way to return to the starting time.

- If everybody was able to use such a time machine and it affects reality, our reality would be constantly changing. Therefore, we should probably also have security measures so that only certain certified people can use the time machine, good rules about when and why to use it, and some system to guarantee that there are no time-related alterations of reality.

Did you come up with more tests than the ones listed? How many were you able to identify? I recommend this exercise with other apps and systems to initiate your thinking about testing and creativity, especially if you don't work directly with testing.

The *top-down* approach should follow the same sort of testing, just in a different (perhaps more chaotic for some people's tastes) order. Therefore, we will not review that approach in this section. However, if you are interested, this is the approach I followed in my blog (see *[2]* in the *Further reading* section).

Now that we have seen an example of how to think about testing for a hypothetical, hard-to-test application, let us see how we would tackle one of the most challenging types of applications – games.

# Challenges in game testing

If you like playing games, testing them might sound like the most ideal, coolest job there could be. However, in reality, if your job is to manually test some screens, levels, or actions over and over again, it may be less fun than you think, and it could even spoil the game for you.

Nonetheless, working in game testing automation is possibly one of the most challenging and interesting areas related to testing (if you don't mind a small game spoiler here and there).

Some of the reasons games are such challenging applications to work with are the uncertainty related to them and the use of specific technologies, such as high-end graphics, cross realities, and artificial intelligence, in their creation.

There are some studies out there that are worth reading in relation to uncertainty. Basically, uncertainty gives a game the *fun factor*. The game Tic-Tac-Toe was fun until people figured out that it is possible to find the optimal strategy to win every time. The moment you realize that the whole game is known and there are no ways to trick or to be tricked, is no level of dexterity or intelligence involved in it, and not even luck (as long as you move first), then the game ceases to be fun and you move on to a different activity.

On the other hand, if a game is easy to automate, it can also be easy for players to cheat (using their own automation), and therefore, it becomes boring. For that reason, game developers try to games in which it is impossible to cheat, but that makes it even more challenging to automatically test them.

There are other things that make games harder to test than other applications – for example, moving items across the screen (background, missiles, sprites, playable items, etc). The objects can also appear in 3D, rotate, or change. Often, when a character is walking or running, even for simpler 2D games, the action requires at least two different sprites representing the character – one with the legs together and the other with the legs in a walking position.

Sometimes, these objects are available in a **document object model** (**DOM**) or some sort of tree, and we can search for some tags in these, as we do for websites. For example, in *Chapter 10*, *Traveling Across Realities*, we saw how to reference the objects from Unity's hierarchy. If more than one sprite represents the same object, we could just iterate over them to make sure we get one or the other.

Other times, we don't have any sort of DOM, tree, or code available for us to retrieve such objects, and we need to make use of other techniques such as AI. We will see more about this particular issue shortly. For more information on AI, see *Chapter 8*, *Artificial Intelligence is the New Intelligence*.

> **Key issue – object identification**
>
> There are some solutions to help us identify objects:
>
> - Use a DOM or a similar tree to find the object or objects
>
> - Keep different screenshots or sprites to identify same elements with different forms
>
> - Use different locators or rules to identify objects
>
> - Use AI to identify objects

Games can frequently combine multiple users to play co-op or against each other. Multiplayer automation is usually harder to achieve than single-player, and for end-to-end tests as true to the real user as possible, this requires some remote communication. We talked about this in *Chapter 5*, *Testing Automation Patterns*, but your system might require a different remote communication than the one exemplified there, so be prepared to get familiar with your specific technology.

> **Key issue – multiplayer**
>
> Use something like RPOM for remote communication testing.

We could find some issues that come from developing in different languages. For example, it is common to find within a game a mix of different screens; one part could be a website that shows the game help, and another could be an API written in JavaScript or similar, with some game logic written in a programming language such as C that allows the game to interact with specific hardware or devices, such as game consoles or controllers (remember that we talked about controllers and signals for VR in *Chapter 10, Traveling Across Realities*).

Be prepared to have more than one framework or type of driver interacting with the other, or to deal with multiple platforms. As an example, we could have a web browser interacting with a phone app to verify a connection (such as the need for two-factor login verification or another sort of inter-device operation).

Let us see how to use multiple drivers in the following Python code (we will just comment on the part of the calls and leave it for you to work on your particular application):

## multiple_drivers.py

```python
from selenium import webdriver
from appium import webdriver as wd
capabilities = {"BROWSER_NAME": "Android",
                "VERSION": "4.4.2",
                "deviceName": "Emulator",
                "platformName":"Android"}
driver1 = webdriver.Chrome()
driver2 = webdriver.Remote("https://127.0.0.1:4723/wd/hub",
                            capabilities)
driver1.get("https://www.packtpub.com/")
driver2.get("https://www.packtpub.com/")
#driver1 initialize action on the browser (example login)
#driver2 confirms action on the app (example, tap 'yes' on
#the popup that appears to confirm login)
#driver1 continues with the action
```

While it is probably not the best idea to use two devices for authentication (we showed other alternatives in *Chapter 4, The Secret Passages of the Test Pyramid – The Top of the Pyramid*), the previous code shows how we could use several drivers to collaborate between them. Please note that each driver gives a degree of instability in a system, as many things could happen with a physical device attached, so try not to use this technique unless it is strictly necessary and there are no other ways available.

That said, there are many other combinations; we could have two phones with two remote drivers, one Android and another iOS, and so on. We could go further and have different drivers for different purposes, on different platforms, including the Unity driver we saw in *Chapter 10, Traveling Across Realities*.

Let us recall how to add yet one more driver – in this case, for Unity control:

## multipleDrivers.py - unity driver

```
# add this to method
dev=UnityEditorWindow()
addr = ('', 5001)
driver_unity = UnityPoco(addr, device=dev)
```

Now, we can use `driver_unity` to interact with our Unity app while we could use other drivers to interact with other apps, just as we saw in the previous example (`multiple_drivers.py`).

Sometimes, having different a driver than an existing one could be our solution, especially when drivers need to be specialized for different systems and/or hardware. In these cases, we need to automate a server to host the driver itself or extend from an existing driver (for example, `remoteWebDriver`) to add our own hardware-specific methods and calls if hosting is not needed.

> **Key issue – existing driver calls are not enough**
>
> The solution is to work on our own drivers or/and extend the calls.

For example, if we want to extend `remoteWebDriver` to add a method, we could do something like this:

## ExtendingDriver.java

```java
public class ExtendingDriver extends remoteWebDriver {
    public void sayHi(){
        System.out.println("hi");
}}
```

To use such a driver, we can just add it to our imports and use it as if it were any other driver. Here is how:

## UsingDriver.java

```java
import com.packtpub.ttm.chapter11.ExtendingDriver.java
public class UsingDriver {
    public static void main(String[] args) {
        ExtendingDriver driver = new ExtendingDriver();
        driver.sayHi();
}}
```

Besides extending our driver, which could be generally good for times when we might need specific driver configurations, we might need to provide a full server to host calls to the driver and decorate them.

Recall the way we start the Appium server and how it receives calls for Appium to execute. In the same way, we could have our own service specific to a controller or device we have that will execute the indicated actions. We will not be adding any more examples in this part of this book to keep things simple, but feel free to check how to create backend applications and servers that will listen to a request and provide a response. You will find the basis for that in *Chapter 3, The Secret Passages of the Test Pyramid – The Middle of the Pyramid*.

Most of the proposed solutions in this section are not specific to games; they could also be beneficial to web applications. However, they are more frequently used for games. For instance, especially in games, more than in many other applications, we can often find randomly timed popups that might cover our objects and obstruct our interactions for automation – for example, a loot box, a reminder for the user, a loading screen, or an updating screen.

Some of these screens might disappear by themselves, but while they are there, they might impede the continuation of the tests and produce some false positives if we are not careful with them.

> **Key issue – the selected frame or window might not be the desired one**
>
> If a screen always goes away after a certain time, set a waiting time (attach a wait to some object we expect it will appear or disappear instead of using a timely wait).
>
> Otherwise, create a screen verifier and make sure you are on the right screen before performing the next action.

Let us see how we can create a screen verifier with Python and the `pyautogui` and `win32gui` *[3]* libraries:

## screenVerifier.py

```python
import pyautogui
all_windows = pyautogui.getAllWindows()
for window in all_windows:
    print(window.title)
# if title is the popup one, close the window
#(window.close()), clicking on some element or waiting for
# it to be gone
```

Now, let us see how to do it with the `win32gui` library instead:

## screenVerifier2.py

```
import win32gui
w=win32gui
w.GetWindowText(w.GetForegroundWindow())
# If the text is not the expected, close window or wait as
# before
```

> **Note**
> If you cannot install `win32gui` directly, try using the `pip install pywin32` command.

You can also use these libraries to switch between windows if you need that functionality.

Another issue is that our test might not always start in the same place; we can be on a different starting screen, for example, if we continue the game instead of starting a new one. The previous screen verifier could be sufficient to indicate this, or maybe we would need to verify based on key objects on the screen.

Loading screens are commonly found in other types of applications as well, so if you only remember one thing from this chapter, let it be the waiting time technique. Please make yourself familiar with the different waiting types (explicit, implicit, etc), be careful with stale object exceptions, and make sure that you understand the differences between methods such as `isEnabled` and `isPresent`, or whatever they are called in your test framework.

There are times, however, when the testing process is yet a bit more challenging. For example, imagine having code that is written in different programming languages altogether, so that instead of a web DOM, you have a different tree (for example, a UIA tree [4]).

> **Key issue – code written in different languages**
> Think about creating tools to retrieve/inspect/dump the different DOM objects, and interact with each object in its particular way.

While working with different trees, consider dumping the contents of the tree (similar to getting the source for a website). In Python, this is done with the `json.dump()` command [5].

To inspect objects in the window, use the `winauto`, `inspect.exe`, and `accessibilityinsights` tools [6].

You might need to be familiar with simulators for each of the systems involved (or even find the need to create one) as well as their communication frameworks and signals.

Besides the software systems, each physical system or platform might come with its specific challenges or designs. Imagine testing a game that needs to transfer from mobile to PC seamlessly. Maybe you need your phone for some features of the game and the PC for others. Or maybe you need to modify your RPOM to include different types of devices for such actions. Although the idea is similar, in this case, the devices might have different drivers or technologies.

> **Key issue – controllers and other peripherical**
> Use or create emulators, and test them on different systems.

While we would love to include an example of controller/hardware automation in this book, it would be so specific that we consider it to be of little help for the general audience. Make sure you learn the specifics of the hardware architecture you need to test. When possible, use software for this, as that would be reusable, and only go to the trouble of learning the lower-level signals to communicate with the devices and language if it is really necessary.

An extra added challenge of automating embedding technologies is that we might need to connect to created servers or open specific communication ports to allow for the interaction with the automation. We then must ensure that those ports are secured during the tests and closed afterward to avoid cheating and malicious intent.

> **Important note**
> If you can automate something, other people will be able to do so. Protect ports and obfuscate information about your automation when needed to protect your system from external hacking.

Localization can come with its own challenge – applications might have a totally different look and feel depending on the region in which they are being sold. For example, when I worked on *Minecraft*, the differences between the western and Chinese versions were quite obvious. In the Chinese version, rather than starting the game directly, your first screen would allow you to chat with friends, purchase objects and tools, exchange items with others, and give you many options that are Western-centric, which would not be as successful or interesting in China.

There is a lot to learn about a country-specific market. What are the users interested in? What do they enjoy purchasing? What makes them excited? What hardware is the most used there? For example, for games, we want to know whether mobiles, PCs, or consoles are dominating the market.

> **Key issue – localization changes the entire system, including hardware**
> The solution is that you may need completely different test systems for different markets.

While games have their own particular challenges, we can still find challenges in them that we find in other applications. For example, elements not being retrievable on a DOM is a common problem in games and other applications. This is particularly challenging when the element is part of a login screen, as it would be a precondition for every test.

> **Key issue – an element not being retrievable**
>
> In order to help with elements not being retrievable, you could try one of these three solutions, or even a combination of them:
>
> a. Try to find an API to test this part or to help with the UI testing.
>
> b. Try keypresses – be careful with keeping devices that use them so that no interruptions or other inputs happen while testing with keypresses.
>
> c. Use object visual recognition (with AI) to find the object.

Each of the previous suggestions has its benefits and disadvantages. For example, with keypresses, we couldn't use the same system to parallelize tests easily, as the keypresses might interfere with each other. There are ways of synchronizing such tests, but even so, it wouldn't be the best solution if you need to parallelize tests on the same system. If we decide on this solution, we should make sure that nothing interacts with the system on which the tests are running, as it could break such tests or cause false positives.

Using visual testing is generally an appealing solution, but it is not always recommended. When every test has to go through a screen such as a login first and the UI changes frequently, it could be more expensive to take screenshots for each new look than manually doing the test. If the fields that we interact with (such as password and username) are not likely to change places and are accessible through keyboard hits, it is probably a better and cheaper idea to use keypresses instead, although using an API for the login, when possible, could be even a better idea.

In the following piece of code, we can see an example of keypresses using C# and Windows forms:

## ExampleOfKeyPresses.cs

```
using forms = System.Windows.Forms;
forms.SendKeys.SendWait("{BKSP}");
forms.SendKeys.SendWait("{TAB}");
forms.SendKeys.SendWait("{END}");
```

The same could be done with the library we saw before and Python, which also allows for more complicated commands, such as pressing *Ctrl + C* to copy in the following way – `pyautogui.hotkey('ctrl', 'c')`. Consider using hotkeys for other actions also, such as switching between tabs.

Keypressing will most surely not be enough for object detection, so we might need to go down the visual recognition route for that task. If that is the case, you might find that the current tools on the market are not sufficient for your case, and you would need to create your own tool for this. The benefit of having your own tool is that you can train it with your specific objects, so the recognition would be tailored and specific. Recognizing objects in games is more challenging than in other apps because of the different backgrounds, sizes, and resolutions.

> **Key issue – different backgrounds, sizes, and resolutions**
> Use AI.

Examples of algorithms that you could look into to identify objects include the following:

- **Template matching algorithm**: This simply traverses an image pixel by pixel and calculates the matching degree with the image to match each pixel

- **A scale-invariant feature transform (SIFT) keypoint detector**: This is more suitable for objects that rotate or that have illumination

- **Advanced ML algorithms using advanced libraries, such as TensorFlow or PyTorch**: These can help you identify 3D or more difficult and changing objects

Be careful with sprites that change – for example, as we mentioned at the beginning of this section, when a character is walking in a video game, there are frequently two versions of the walk. Depending on how such character was created, it could be part of the same object or a different one altogether. You might need to add identification for any of the possible positions in order to retrieve the object.

We are purposely not including such algorithms in this book, as it would make things too AI-specific if we did and would require specific chapters for each of them. Feel free to research more on these and other algorithms to fit your needs.

Another problem with games occurs while playing them. To get the most similar experience to a user, you might need to add some AI to the game, although it is, so far, less expensive to do a test manually. For example, Microsoft's project, Malmo, attempts to solve different challenges for the game *Minecraft*, but it is still far from a full game experience. However, other games have successfully found solutions by using AI, especially by using reinforcement algorithms, such as the game *Space Invaders*. If you ever need a library to solve games such as that one, check out Gym. We covered the basics for reinforcement algorithm using Gym library in *Chapter 8, Artificial intelligence is the New Intelligence*.

> **Key issue – automating a full game experience might require intelligence to win the game**
> Use AI, reinforcement algorithms, or manual tests.

Alternatively, you could try to expand and check every action. This is also commonly done in other applications, with what is known as a crawler or exploratory testing tool, but, as you can imagine, this is harder to create for games. The goal of a crawler is to discover functionality by executing all possible actions. We saw an example of this in *Chapter 7, Mathematics and Algorithms in Testing*.

The benefit of using a crawler is that you do not need to write code explicitly, and it has better tolerance to game changes, but creating one for a game is much more difficult than creating one for web applications. Examples of difficulties when creating a game crawler are as follows:

- Finding the intractable elements might not be as easy in a game as when we have a clear tree or DOM where objects are listed. We have already explored what to do about this.

- Understanding where each "screen" is also more difficult. Sometimes, the game might have reached a checkpoint that looks exactly like another section of the game but requires a different interaction. We need to make sure we are in the right screen, for example, checking the expected elements for that screen or logs.

- Some functionalities might come from a different action than pressing a button, such as swiping in your phone, or an external device command (such as console controllers or VR handsets). So make sure you include these actions in your crawler.

> **Key issue – automating exploratory testing**
>
> Consider creating a crawler. This might be a bit more challenging when it comes to games (but we have just provided some solutions to solve some of these challenges).

The most recent problem I have faced when dealing with games is automating cross reality. We discussed this in depth in *Chapter 10, Travelling Across Realities*.

Cross reality might be challenging to automate in such a way that simulates the experience of a user – for example, some objects might not be directly visible to the user, and when automating the rotation of the user, we need to make it work as slowly and smoothly as a user would rotate in real life.

Besides the automation, we would also need to take into account some other issues that might appear specifically in cross-reality environments, such as sound, textures, and others explained in *Chapter 10, Travelling Across Realities*.

> **Key issue – automating in cross reality**
>
> We explored solutions in *Chapter 10, Travelling Across Realities*.

Games are fun but also challenging to automate. There are many different technologies that need to be used, depending on the use case, so having an open mind and willingness to learn new things is a must in order to succeed.

AI, machine learning, and computer vision can help us locate difficult objects and even create automated testing. However, sometimes, these solutions might not be ideal for your particular system.

Automation for cross reality is achievable, but it comes with its perks and unique challenges.

Specific software or hardware also requires its own particular frameworks, drivers, and automation systems in general.

In the next section, we will explore what to do about some other difficult-to-automate applications.

## Automating the *impossible*

There are some times that we are told that automating this and that is *impossible*. I personally relish this challenge. Consider that we have all these inventions and applications nowadays that years ago would have seemed impossible. We have gotten so used to some of them, such as mobile phones that allow us to connect with anybody anywhere in the world, even with real-time video, which sounded impossible many years ago.

Using AI, some apps seem to be capable of foretelling behaviors, virtual reality combines with our reality to create amazing and entertaining experiences, and the internet of things allows wristbands and watches to know whether you slept well enough, how many calories you have burned today, or even if you have washed your hands thoroughly. This all seemed like science-fiction material just a few years ago.

We are starting to see AI helping both development and testing.

Imagine that you have found something challenging and interesting to automate, and you have the feeling that this thing is actually within your capabilities to automate. It's great to dream, but we also have to keep our feet on the ground, so it's important to keep in mind that automation is all about keeping balance. The rule for this is quite simple, and we have already highlighted it a few times in this book.

> **Rule for automation**
>
> If your automation is going to save more time (to you, to others, to the community, or to your company) than you would spend trying to automate the manual efforts, then the challenge is worth a try.

The key here resides in estimating correctly how long it is going to take you to do this automation. Estimations are not easy to do, especially when you are dealing with new, unfamiliar technologies or out-of-the-box solutions. Unless you are totally clear about how long it will take you to automate something, give or take some unexpected issues, you should take your time during the research phase. Use the analogy approach, as we saw in the first section. Ask mentors or other experts in your organization to review your estimations, and ask for help if needed. We will talk more about mentors in the next chapter, *Chapter 12, Taking Your Testing to the Next Level.*

If you feel you are going to take longer to automate something than the time it will take to do the necessary manual testing (say, for example, it is a feature that will be deployed/updated very little), it is okay to drop the automation at any given point, even after the research phase. Alternatively, you can keep it as a side project if you wish, especially if you feel it would help at another point, but do not let it eat into the time spent on your actual achievable and higher-priority tasks.

So far, we have reviewed how to tackle hard-to-test applications, impossible automation, games, and even time machines. In the next section, we will see how to apply automation to everything else so that it saves us valuable time.

## Automating for a higher purpose

Time is one of the most valuable currencies we possess. Yet we keep wasting it and our energies doing tiresome and repetitive tasks. This is why I love automation and creating tools, not only for testing but also to help speed up the entire development process and, really, for everything in life.

Next, we will discuss a number of things you could automate, by coding or with tools, and some hints on how to do so. Some of the things in the list might be obvious to you, but many people tend to forget about them, so use this as a refresher if you already knew about them.

While it takes no computing effort to deal with it, one thing that people tend to forget is that they can automate email sorting. Creating rules to automatically sort out your email in folders might seem tiresome, but it will save you time in the long run. The steps on how to do that are different depending on the email provider that you use, so we will not review this here. However, remember to check all of your folders, or you might miss important emails if they go automatically to another folder that you tend to ignore.

> **Remember to automate**
> Email sorting – create rules to sort your email.

You can create aliases for the most used commands in your preferred terminal. For example, you can create an alias to open a very long folder name, with the following commands:

### Alias creation windows

```
doskey cdLong=cd "longpath"
```

### Alias creation unix

```
alias cdLong='cd "longpath"'
```

When you use cdLong, it will take you to a specific path instead of you having to type the path name every time. There surely are other long commands that you manually type in; think about them and create an alias for them in your computer or environment to save yourself some time in the long term.

> **Remember to automate**
>
> Automate aliases for long commands.

Furthermore, you can create small batches or scripts to automate other long or frequent tasks. This is also useful to do if you need to transfer your work to a new system so that the initial setup is done automatically. Going a step further, if you need to set up something such as a test environment, this can also be automated. This links back to what we learned about Docker and automation files in *Chapter 6, Continuous Testing – CI/CD and Other DevOps Concepts You Should Know*.

> **Remember to automate**
>
> Automate frequent tasks in batches or installation steps on Docker or automation files.

Another commonly frequent task could be sending an email, for example, with a dashboard of test results, automation state, finished processes, or alerts when an important test has failed. Let us see how to send an email in C# with Gmail. For simplicity, we will use a parameter that tells us the number of failed tests, but consider how you can retrieve this from a document:

## EmailSending.cs

```
private static void sendGmail(int failedTests) {
    String userName = "yourEmail";
    String password = "yourPassword";
    String toEmail = "toEmail";
    String fromEmail = "yourEmail"; // likely same as userName
    SmtpClient client = new SmtpClient("smtp.gmail.com", 587);
    MailMessage message = new MailMessage(
        new MailAddress(fromEmail, "Your name"),
        new MailAddress(toEmail));
    message.IsBodyHtml = true;
    message.Subject = "We found some failed tests";
    message.Body = "<html><head></head><body><p>Hi!</p><br>
                    <p><font size=\"3\" color=\"red\">We found "
                    + failtedTests +" failed tests. Please
                    verify.</font></p></body></html>";
    client.EnableSsl = true;
```

```
    client.Credentials = new NetworkCredential(userName,
                                                password);

    client.Send(message);  }
```

Note that you will have to set your Gmail account to be able to access it from the code in such a way by going to `https://myaccount.google.com/lesssecureapps`, and this cannot be done if you have two-factor authentication enabled. Therefore, using the preceding code to send an email, although possible, is highly insecure, and you should research more secure options to automate it. Here, we just wanted to showcase how it could be done easily. *If you try this code as is, make sure you disable this setting afterward.*

To make the preceding code secure, you likely will need to use an OAuth mechanism, such as the one we reviewed in *Chapter 4, The Secret Passages of the Test Pyramid – The Top of the Pyramid.*

In a similar way, we could automate the reception of an email. For repetitive emails or when some action has to be taken once an email is received, this can be useful. For this, besides the code to receive an email, you should also find information on the email itself, which could be as easy as finding a string or something more challenging. However, if you end up on a system that needs to take actions based on emails, maybe you should consider setting up some queue or notification system, such as the one we reviewed as part of *Chapter 3, The Secret Passages of the Test Pyramid – The Middle of the Pyramid.*

If the issue that is happening needs faster attention, maybe you should try sending an SMS or even making a call. For a DevOps system, it is common to find automatic calls to a person *on-call*, stating the issue for high-severity tickets.

> **Remember to automate**
> Automate notifications, over emails, SMS, Slack, WhatsApp, and even calls.

One tool that you can use to automate such things is Twilio (note that it is not free, but you can test it out with the trial version). Feel free to find alternatives or even use a GSM modem if you cannot trust third parties.

In order to send a text or perform an automatic call with Twilio, you need to create an account and then go to `twilio.com/console` to find your account SID and token, and check your assigned `fromPhone` parameter for the operations. The following code is also insecure; see *[7]* for options to make it secure. The following code is an example of how to automate a call with Twilio:

## TwilioCallExample.cs

```
private static void callOnCall(int issueNumber, String
accountSid, String authToken, String toPhone, String fromPhone)
{
    TwilioClient.Init(accountSid, authToken);
```

```
CallResource.Create(twiml: "<Response><Say>Dear person
on Call! We have found an issue "+ issueNumber +"
that needs immediate attention. Some other instructions...
</Say></Response>",
to: new Twilio.Types.PhoneNumber(toPhone),
from: new Twilio.Types.PhoneNumber(fromPhone));
}
```

Here, fromPhone and toPhone should be written as follows – +countryPrefixPhoneNumber. For more information, visit *[8]*.

As before, we can automate a reaction to the reception of an SMS or even a call.

Similarly, we can use WhatsApp, Slack, WeChat, or even Skype for our notifications. For example, we can use Twilio to send a WhatsApp message, including media. Let us see that in the following code:

## TwilioWhatsAppExample.cs

```
private static void whatsAppMessage(Uri[] mediaUrl, String
accountSid, String authToken, String toPhone, String fromPhone)
{
    TwilioClient.Init(accountSid, authToken);
    MessageResource.Create(mediaUrl: mediaUrl,
            from: new Twilio.Types.PhoneNumber(fromPhone),
            to: new Twilio.Types.PhoneNumber(toPhone)
        );
}
```

Here, fromPhone and toPhone are as follows – whatsapp:+countryPrefixPhoneNumber. For more information, visit *[9]*.

As before, Twilio is not your only option; if you want to go lower level, you can check WhatsApp's API directly at *[10]*.

One thing that I have been asked a lot in conferences and online is for help to read documents. It is quite easy to find particular fields on a text or cell document, or to analyze the output of log files. We saw how to do this easily with a regular expression in Python in *Chapter 6, Continuous Testing – CI/CD and Other DevOps Concepts You Should Know*. Let us review how to do that (note that the regular expression is very simple in this case, but we could find more complicated expressions) in the following code:

## FindTextPython.py

```
import re
file_to_analyse= open("fileName.txt", "r")
found_string = re.findall("stringToFind",
                          file_to_analyse.read())
```

There are some libraries that can also help you read PDF documents with Python (and other languages) – for example, Tika *[11]*.

**CSV** (**comma-separated value**) documents are very interesting too. We briefly saw how to use the pandas library to read these documents in *Chapter 8*, *Artificial Intelligence is the New Intelligence*, which makes everything much easier with these documents. Keep in mind that most cell documents can be transformed into CSV documents, so this automation is particularly powerful.

Let us remember how we can read a CSV document into a dictionary:

## ReadingPandas.py

```
import pandas as pd
dataFrame = pd.read_csv('doc.csv', index_col=0)
dataDictionary = dataFrame.transpose().to_dict()
```

Besides reading the documents, you can also create them, add or edit data (as we saw in *Chapter 8*, *Artificial Intelligence is the New Intelligence*, when we curated the data), or even style them (we saw how to style the HTML contents of an email message before).

> Remember to automate
> **Create, Read, Update and Delete** (**CRUD**), style and organize documents.

We can create some very nice reports using this technique. Let's revisit the JavaScript library we saw when we created dashboards in *Chapter 7*, *Mathematics and Algorithms in Testing*.

Let us put some of it together:

## TestReporting.py

```
import re
import subprocess
result_file = open("result_file.txt", "w")
run_test_process = subprocess.run(routeToRunTestsFile,
```

```
                                            stdout= result_file)
file_object  = open("result_file.txt", "r")
failed_number = re.findall("Failed: .", file_object.read())
iflen(failed_number) >0:
    failed_tests = failed_number[0][8:]
    subprocess.run([routeToSendEmail", failed_tests])
```

This code will run the process to run files – for example, the tests created with `vstest` tool we saw in *Chapter 6, Continuous Testing – CI/CD and Other DevOps Concepts You Should Know*, it places the output of test execution into a text file, then opens that file for reading, tries to find the `failed` keyword, and, if there are failures, calls the process to send an email that we covered before. This is basically an extension of the `LogAnalyser.py` program in *Chapter 6* that also calls the other programs needed. See the last section of this chapter, *Architecting a system*, for more information about deciding between the system of *Chapter 6* and this program and why this solution might not be secure.

Organizing documents can also be interesting to automate. For example, in *Chapter 5, Testing Automation Patterns*, we saw different POM patterns, including record and playback (EPOM). With this pattern, it would be useful to remove old files, such as old screenshots that are not needed for the code anymore. Let us see how to rename a list of images, considering that images are all that there is in a path (otherwise, we would have to look at their extension):

## imageHandling.py

```
import os
path = r"path_to_images"
list_of_images = os.listdir(path)
def rename_image(self):
  count = 0
  for image_to_find in list_of_images:
        os.rename(path+image_to_find,
          f"{path}{count}image_to_find.getName()")
        count = count + 1
```

If we want to remove unused images, we would first have to iterate through the code to find out which images are used and add them to a list, set, or dictionary. Then, we would iterate through the images in the directory we want to clear and remove the ones that are not part of the dictionary or set:

## limageHandling.py

```
## find used images and save them into set "img_set"
## iterate through all images if not in used --> remove
def remove_unused_image(self):
    for img in list_of_images:
        if img not in img_set:
            os.remove(img)
```

Finally, remember that even your home can be automated to provide security and comfort, be environmentally friendly, and save you time. Let us see some examples:

- You could power on your computer and your coffee machine or kettle with a smart switch. For example, if you are *on-call* and you do get a call, this could happen automatically so that you can focus on washing your face to wake up while your computer turns on and your beverage is ready.

- You could automate the lights in your house to turn on at a certain time and turn off at another time (using, for example, a socket adapter with a timer). This way, you do not need to double-check you have turned them off before you head to bed at night.

You can also have cleaning bots, smart cooking devices, temperature control, and even automated blinds or curtains. With automation, you can let your imagination flow!

> **Remember to automate**
> Automate your home and environment to give yourself some extra time.

Remember the automation steps that we mentioned in this book:

1. Identify repetitive tasks.
2. Write code that will do these tasks for you (or get a smart socket or timed socket for this).
3. Identify when the code needs to be executed (or when you need them to turn on or off).
4. Execute the code or automate the execution (or the preceding program/configure). Identify success measures (verify if there is something you need to do to improve this automation or further on it).

Furthermore, why not give someone you appreciate some automation? As we mentioned before, time is one of our most valuable resources, so giving automation is as powerful as giving time. I've seen many colleagues creating small scripts here and there to save themselves some time, but they rarely share those scripts with others, so do share your automations, build on them, and share them again! Let us build a culture of helping everyone save some time!

## Record and playback tools

When dealing with automation, we have many programs and tools available to us. Some of these allow us to record actions, save those actions as generated code, and play them back on demand. We mentioned them previously when we talked about EPOM in *Chapter 5, Test Automation Patterns*. They can speed up a process in comparison to writing all the related code to automate the necessary actions. However, there is a reason many companies prefer to write code than use such tools.

While the *generation of code is faster*, especially when it comes to repetitive pieces of code such as the skeleton of the program (imports, function calling, method headers, keywords, and others), the *maintainability* of such code is generally *harder*. When something changes in a UI, it is generally faster to re-record an entire test than to try to find out what has changed to fix the generated code.

The code that gets generated is also *harder to scale* if we want to reuse such code in different platforms, devices, regions, and so on. Let us see an example of some generated code with the Selenium IDE [12] for clicking to navigate to a login screen (the rest of the login process is omitted to simplify the code):

### test.side

```
{
  "id": "7ed170d1-1285-429e-97b9-d83d3cf3a1b7",
  "version": "2.0",
  "name": "test",
  "url": "https://www.packtpub.com",
  "tests": [{
    "id": "98adb08f-52e7-48d7-a795-a8fd0ac46331",
    "name": "Untitled",
    "commands": [{
      "id": "facef1c1-1543-427e-b23a-55ae3bb66b7b",
      "comment": "",
      "command": "open",
      "target": "/",
      "targets": [],
      "value": ""
    }, {
      "id": "ef8a4083-27e6-4bf4-b009-d3bf1e4519ea",
      "comment": "",
      "command": "setWindowSize",
      "target": "1296x696",
      "targets": [],
      "value": ""
```

```
  }, {
    "id": "1b005e96-7a32-48c1-ae80-2b72580f06ae",
    "comment": "",
    "command": "click",
    "target": "id=__BVID__217__BV_toggle_",
    "targets": [
      ["id=__BVID__217__BV_toggle_", "id"],
      ["css=#\\__BVID__217__BV_toggle_", "css:finder"],
      ["xpath=//a[@id='__BVID__217__BV_toggle_']",
       "xpath:attributes"],
      ["xpath=//li[@id='__BVID__217']/a",
       "xpath:idRelative"],
      ["xpath=(//a[contains(@href, '#')])[3]",
       "xpath:href"],
      ["xpath=//div/ul/li[4]/a", "xpath:position"]
    ],
    "value": ""
  }, {
    "id": "fdb485b3-28c1-42ee-99cf-df2d316f544f",
    "comment": "",
    "command": "click",
    "target": "id=__BVID__217__BV_toggle_",
    "targets": [
      ["id=__BVID__217__BV_toggle_", "id"],
      ["css=#\\__BVID__217__BV_toggle_", "css:finder"],
      ["xpath=//a[@id='__BVID__217__BV_toggle_']",
       "xpath:attributes"],
      ["xpath=//li[@id='__BVID__217']/a",
       "xpath:idRelative"],
      ["xpath=(//a[contains(@href, '#')])[3]",
       "xpath:href"],
      ["xpath=//div/ul/li[4]/a", "xpath:position"]
    ],
    "value": ""
  }
...
```

You can see the rest of the document on GitHub: `https://github.com/PacktPublishing/How-to-Test-a-Time-Machine/tree/main/Chapter11/test.side`

If we were to try to repeat this code in other locales, we would have to record the same entire test for every locale. If we wanted to modify it so that we can reuse the automated code in other locales, we would have to search around for the parts that are locale-exclusive (the navigation part and the clicks).

Furthermore, we would need to validate that the rest of the elements are still found in each of the locales (which might not be the case if the recording tool has recorded things that could easily change, such as names). In the end, the entire process can take us much longer than if we were to automate it from zero, and also uses more space than if we did it right and utilized a data source or a loop, instead of a file, for each locale.

With record-and-playback tools, we may want to create another test that performs common steps with the previous one. In our example, we login the user, so if we need the user to be logged in with the recording tool, which means that all recordings would have to start by login the user in, and we would need to record the same steps for the login repeatedly. Therefore, if most of our test cases have a good deal of repetitive actions in common, it would be better to automate writing code than using a recoding-and-playback tool, as this allows you to extract and call the repetitive actions as a function or method.

Although most record-and-playback tools will allow (or won't forbid) you to edit generated code so that you can copy and paste the required commands, it will end up *costing more space* to save the generated code than if we had written a function. Even if we could edit the generated code and create a function for a piece of generated code, the rest of the recording could be just as complicated and time-consuming because to reach that part of the recording, you would have to have performed that action every time anyway. Therefore, record-and-playback tools have *low reusability* for repetitive common actions.

Record-and-playback tools that work with screenshots have their own set of issues. Maintaining the screenshots taken to find objects can be time-consuming, and not maintaining them will lead to *testing debris* and cost computer space. Some tools can help you clean up screenshots that are not used in code, but if you share the same folder for several classes, it could remove the screenshots that are used by the other classes too, breaking the rest of the automation. Be careful with this and create your own functions to clean up space, as we saw in the `imageHandling.py` program, if you use a record-and-playback tool that work with screenshots.

If harder maintainability, reliability, scalability, increased test debris, and decreased reusability are the inconveniences of using record-and-playback tools, why should we bother to build or use them at all? The answer lies in their benefits:

- They have a faster turnaround, which makes them ideal, for example, for a quick check or while deciding on frameworks.

- They are usable by different team members with mixed skills, allowing everyone in a team to create automation for testing or other purposes.

- They allow for other ways of finding objects other than using DOM. This is very useful for hard-to-test applications such as games and XR, especially if we are using screenshots. When we create a screenshot to recognize an object, we should make sure that it works well in the automation; we might need to take a different screenshot with a different size or of different sections of the item to be found. If we do this within a tool, we can take it at the time of the test record (so we go through it only once), validate it, and edit it there. Otherwise, we need to reach the point that needs the screenshot, take it, save it, edit it, remove it, and save it again, which is very manual.

Last, but not least, we can use record-and-playback tools as an aid in our automation, to find precise objects and see their attributes, reutilize some of the generated code, verify the functionality under test, and so on.

Even for a tool such as the Selenium IDE, which saves objects in a dictionary structure without any specific programming language, we could turn such a structure into reusable code for our programs in an easier way (I will leave you to figure this one out as an exercise).

Let us now review a few final things that were not covered in the previous chapters and are important for automation too.

## Last-minute tips and best practices for your automation

I am sure I will forget to mention many important things in this book (besides the ones I might not be aware of myself). Before moving on to how to architect a system and put it all together, I think it's important to mention some best practices.

A common question asked in the testing world is how to test the uploading and downloading of files. When your application needs to download or upload a file, skip that bit during testing, and test all the way to before and after the action. You can use an API call to retrieve the file and then check for specific values within the file. You can check that a form shows an error message if a file is not attached. However, do not waste time making sure that the file is downloaded, as that is up to the browser to handle. If the browser does not download the file, the problem lies beyond your app.

We could say a similar thing about third-party applications; it is good to test the integration with an app, but do not test functionality that you cannot influence if you find a defect on it. Ask yourself: "*If this fails, is it up to us (the developers, the company, or even the team) to fix it?*" If the answer is *yes*, you need to test it; if the answer is *no*, just test that your application responds as expected if such a thing fails.

---

Tip

Test your application and integrations; do not test third-party apps or browsers.

---

When dealing with web applications, it is common to use the programming language that is more comfortable for us. However, **JavaScript (JS)** is very powerful for handling websites, and we should remember that many other languages will allow us to use JS executors to add small JS snapshots, which can help us take advantage of such power.

> **Tip**
> Don't underestimate the power of a JS executor.

The next tip is something we have been saying throughout the book frequently; nonetheless, I think it is very important to reiterate. Avoid using too much UI. We should just use UI when we try to mimic the action of users, but for the actions that we have already tested somewhere else or that are not relevant to our applications; we should always try to use some shortcut or an API.

> **Tip**
> Use an API whenever possible.

We have mentioned several times throughout the book, and in this chapter specifically, what we should and should not automate.

> **Tip**
> Automate everything repeatable that will save you time.

You will be surprised by the number of things that you can easily understand collaterally out of gaining an understanding of something else, and that will give you the ability to relate different systems and inspire your creativity.

> **Tip**
> Invest in learning and your own education.

Make sure that whatever you work on will outlive you at your team or company, so that it will work well even when you are not on charge of it anymore. If a test is always functioning wrongly and failing even when there are no bugs, nobody will trust it – just like the folk tale of *the boy who cried "Wolf!"*.

> **Tip**
> Create a reliable test framework and tests.

Now that we have reviewed many tricks, tips, and exercises that will help us have a high-quality testing system, let us now see how to put it all together so that we can design such a testing system, and see how the test architecture connects the different systems.

## Architecting a system

We have seen some of the options we have whe building automation for our system, what is left is how to put our system together, as this book is about test architecture, not only its components.

There are many ways we could automate our system and many choices we could make. A good system architecture will allow you to replace components with others when needed, in such a way that the system will still work.

For example, in this chapter, we have seen how to call different programs within another program in the testReporting.py code. However, in *Chapter 6, Continuous Testing – CI/CD and Other DevOps Concepts You Should Know*, we did something similar as a part of a CI/CD system in the logAnalizer.py code. How shall we decide between the two ways?

For simple applications or a quick turnaround, we could go for the solution in this chapter. However, if some of the programs that are called with subprocesses were to break (or someone deliberately breaks them or adds mean code), the entire system collapses. Therefore, the approach shown in *Chapter 6, Continuous Testing – CI/CD and Other DevOps Concepts You Should Know*, is more suitable in this case.

An even better test architecture would allow us to replace any of these components with another one anytime. For example, we may want to change the way we call the test cases from vstest to some other tool or language, or we may decide to take a different or additional action when we find a failure.

Therefore, we could have a system in which the test execution can be done in any language and will notify us through a queue or some other messaging service when it has finished, alongside storing the results somewhere, as it could be a text file or some cloud service that allows for storage (such as an S3 bucket).

Then, the results verification part of the system will grab that file (the address or name can be statically shared or can be given by the notification system or another parameter or configuration file) and will verify the results.

Finally, we can have a part in a system that can update or create reporting. This could be, for example, a static or dynamic dashboard or even an email message. It could alert the person *on-call* whether there was a high-severity incident or a high-priority test failing (through WhatsApp, a call, text, etc.), and/or create a bug (in Jira, TestRail, etc.).

See the next generic diagram to exemplify this design:

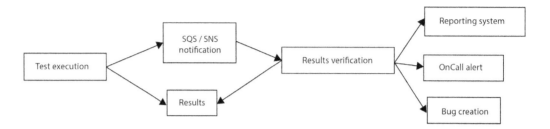

Figure 11.1 – An example testing system architecture

If we were to decide that we do not want to alert an *on-call* person about failing test cases, we could take that bit out. Or, if we decide that we want to add additional actions after seeing the results, we could simply add another block. Imagine, for example, that we have a yellow light in the office to mark when a big issue happens; we could add a box after results verification to turn such a light on.

For this system, the stability of test cases is a given; otherwise, we could be alerting people about false positives, so be careful with this part. Most systems will proceed to call the *on-call* person only when a high number of issues have been reported in a short period of time, and they generally come from metrics in the system in productions and user reports, rather than automation itself (that would be a different part of the system that does not require alerting).

My personal preference would be to get a Slack notification when the system has finished, so I would add the Slack note after test execution and remove the *on-call* alert (at least from this part of the system).

Another possibility is to add another verifier after the bug creation, and when a number of opened bugs reaches a level, then make the call to alert the *on-call* person, so that not only this system will be feeding into it. It could run right after every bug is created or continuously search for high bug numbers. Note that this system might just run during working hours, so this *on-call* person is not necessarily going to get a call after hours for a high number of bugs. Be sure you specify what should be the role of the person or people that are *on-call* if you include them in your architecture.

In a similar manner, we might want to update the dashboard when all different sorts of testing have been performed. We could have notifications coming from several systems and update the dashboard with the new information, either when all of the notifications have finished or on a regular basis – for example, every day.

Taking all of that into account, see the following diagram portraying it:

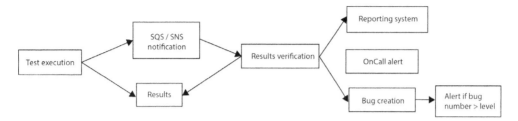

Figure 11.2 – A better example of a testing system architecture

A proper system design would specify each of the elements on the diagram, expanding on them in a document.

Some questions to ask when creating such systems that can help you understand your needs better are as follows:

- What is the expected number of people using the system?
- Would we need to escalate at some point? When?
- Is the system scalable, maintainable, secure, and reliable?

Make sure you ask yourself (and others) as many of these questions as possible to get to the best out of the design.

Note that some parts or blocks of the system might be done with different third-party apps. The important thing is that our design describes the interaction between the different components.

You can have more than one test architecture diagram, balancing the pros and cons of each of them, and present them to your team, customers (who could likely be other team members), and management to help you reach a conclusion about which one of the systems to go forward with.

Most times, architecting a design is quite an abstract thing to do, especially when you are new to a team or a feature and there are a lot of unknown factors for you. The good thing is that you can always iterate through the architecture and keep growing and learning. That said, getting as much information as possible at the beginning is key for a great design, and it takes a lot of practice to get there, so use any opportunity that you can (while writing any tool) to try to design systems, and read about system design (even for those not only specifically for testing). Several companies, including big companies, provide examples of designs they have used when creating different features or their systems, which make a great read if you want to learn more about it.

Keep in mind the entire development system. Firstly, there will be more tests than the ones we covered here, which were UI tests. All the different tests will have to run at different points of the system, and integrate the results, metrics, and alerts accordingly.

Secondly, and in relation to the previous point, you might or might not be part of a CI/CD environment, and we could go further and add the different environments and deployments, post verifications and data analysis, alerting, metrics, load balancing, caching, debugging, and so on. If you are not part of such designs, still make sure you review them and agree on them. At the end of the day, if you are reading this, chances are you are passionate about quality, and quality is not just about testing and test cases. You should have something to say about the quality of the system and whether all possible scenarios are covered.

## Summary

In this chapter, we have seen some tips and tricks to test challenging applications, covering games, including difficult ones, non-testing automation, and even what to do with impossible-to-test applications.

We started the chapter by reviewing how to test a time machine (which gives this book its title). We saw how to approach the testing of a system in an ordered way and how to think about all levels of testing and all the testing types that we could implement in our system. This could be extrapolated to other apps, the time machine being an example.

Then, we reviewed a list of different issues that hard-to-test applications can have and some potential solutions to those issues. We continued the chapter by recalling things that we could automate that we tend to forget about, including record-and-playback tools and their advantages and disadvantages, followed by some general automation tips.

We finished the chapter by talking about test architecture (which also gives this book its title) and how to design a testing system.

This chapter is full of practical examples, and we hope they can help you practice and come up with interesting and new approaches to the issues covered and beyond, as well as difficult-to-test applications. Make sure you spend some time practicing and playing around with the technologies that we have covered here.

In the next chapter, we will see how to take your testing to the next level by challenging your career and helping others, along with some tips on how to grow in the testing field.

## Further reading

- [1] The Apollo 11 code was uploaded here: `https://github.com/chrislgarry/Apollo-11`.

- [2] My blog is `https://thetestlynx.wordpress.com`; you can check the past posts section to find the one about the test machine.

- [3] For more information about the `pyautogui` library and how to control the mouse and keyboard with Python, see `https://pyautogui.readthedocs.io/en/latest/`; for the `win32gui` library, visit `http://timgolden.me.uk/pywin32-docs/win32gui.html`.

- [4] More information about the UIA tree is available here: `https://learn.microsoft.com/en-us/dotnet/framework/ui-automation/ui-automation-tree-overview`.

- [5] Working with JSON objects can make things easier; you can find out more about the JSON library for Python here: `https://docs.python.org/3/library/json.html`.

- [6] WinAuto is a bit old now: `https://docs.microsoft.com/en-us/windows/win32/winauto/inspect-objects`. Accessibility insights is the recommended tool to inspect Windows elements at the time of writing: `https://accessibilityinsights.io/docs/windows/overview/`. See also the `inspect.exe` tool: `https://learn.microsoft.com/en-us/previous-versions/troubleshoot/winautomation/process-development-tips/ui-automation/using-inspect-exe-to-access-ui-elements-winautomation-is-not-able-to-see`.

- [7] Visit `https://twil.io/secure` to make your Twilio application secure.

- [8] Visit `https://twil.io` for more information about creating Twilio applications.

- [9] See more about sending and receiving WhatsApp messages with Twilio at the following link: `https://www.twilio.com/docs/whatsapp/tutorial/send-and-receive-media-messages-whatsapp-csharp-aspnet`.

- [10] See Whatsapp's API at `.https://business.whatsapp.com/developers/developer-hub`.

- [11] There's more about the Tika library for Python here: `https://github.com/chrismattmann/tika-python`.

- [12] Selenium IDE for recording and playback: `https://www.selenium.dev/selenium-ide/`.

# 12
# Taking Your Testing to the Next Level

We began our journey in this book by trying to identify where we are in terms of quality. Throughout, we have seen different levels of testing and technical tips and techniques to learn more about each of them. In this chapter, we will review softer ways of taking your testing to the next level, including finding your style and interests, figuring out what skills you will need if you want to make a professional move or transition, how to get inspiration, and how to inspire others and help them grow as well.

Although this chapter has been written for people that work in quality-related positions, many of the topics could be applicable and of use to people that work in other parts of the software development life cycle, and especially, for movement across the roles.

In this chapter, we will cover the following main topics:

- Challenging your career:

  - Finding hot topics

  - Finding your style – horizontal versus vertical knowledge

  - Finding time to work on the hot topics

- Quality growth:

  - Helping people in your team grow

  - Transitions and career opportunities

  - Impostor syndrome

- Closing thoughts:

  - Finding your why

  - Ethics in testing

  - The future of testing

By the end of this chapter, you will be ready to start challenging your career and also help others challenge their careers. As you do this, the collective productivity and inspiration will also increase, giving you yet more energy. Therefore, remember this chapter when you are running low on energy or inspiration.

## Technical requirements

None! This is a closing chapter where I intend to get you excited about automation, quality, working on your career, and sharing success with others.

## Challenging your career

It is easy to follow the known path or get stuck in repetitive test automation. It takes courage to try new things and to get out of your comfort zone, especially when so many things are constantly being created. By the time I wrote this book, at least 10 other people in my network had published theirs.

Later in this section, we will talk about *hot topics*. I have found people that are after the latest hot topic in the industry because they just want to create something big or want to get their names out there, look for a promotion, or make money. They basically start with the solution or technology and try to find problems that justify them working on it. I don't believe this is the correct approach to get any of that. Instead of focusing on the results and benefits you will get from doing that, try to think about the journey, the learnings, and how the results will help you and others. Start by finding a problem you might already have, then work on getting a solution, even if it is not perfect, then check whether you can improve it.

In my case, I have a particularly curious mind. I want to understand how everything works. I tend to prefer the bigger picture to the details because I can move faster to the next thing there is to learn. I recognize that not all minds work in this way, but let me share my tricks during this section in case you find any of them useful too.

We will begin the section by talking in detail about how to utilize the knowledge gained through different mediums and different platforms to get inspiration and help you come up with new ideas:

- **Books**: Books are a wonderful resource for obtaining knowledge, besides creating inspiration. We could also include articles and videos in this category.

  You are reading a book right now, so I guess you are fine with this one. However, let me give you a tip that works for me: if you want to commit yourself to reading, try setting a reading challenge or joining or creating a reading club. Keeping notes and tracking the books I read helps me too so that the books I read are not forgotten, and I can refer to the notes when needed. If you are competitive, try to compete with yourself rather than others and keep your focus on quality over quantity.

- **Goals**: Make sure you work on planning and writing down your goals, achievements, and values. There are plenty of articles and books with great advice on how to do this. It helps me to keep my goals nearby so I read them frequently. This helps me focus on the things that are most important to me.

  Be specific and realistic about your goals, and make sure you can measure your progress with them. I am not generally very good at doing this, so it is "one of my goals" to start doing more of this in particular.

  Try other activities and keep balanced to avoid burning out so that your goals feel exciting. You can also plan your rest periods to make sure you don't overdo it, although be careful with this; it happens to me that if I plan my breaks, they also feel like work somehow. I have felt burnt out so many times that I have to make sure I set reminders to myself, as once you get burned out, it is even harder to get back on track.

- **Commitments**: Having a commitment can help you achieve a goal. For example, participating in meetings or conferences helps me to push myself to achieve a project to present on them. Joining a course can also be motivating, as it pushes you to work on exercises and lectures and will have deadlines you will need to meet, especially when the course costs money, as it would be a waste not to finish it. You can also try writing a blog or magazine, participating in a hackathon, or podcasts and so on.

  A programming group or project could also be an interesting commitment to challenge yourself, for example, working on open source projects, apps, or games. This could also be something from your own company; if it can bring your company revenue, it is likely to be OK with you working on side projects. If you don't find any that suits you, start one.

- **Traveling**: Getting out of your environment can inspire you to create different things. I found that attending conferences is a great way of seeing different places and getting to know other people working on different things that might inspire you.

  If you are not in the position of traveling far, try getting some fresh air and being in contact with nature often. Meditating or praying in nature is beneficial too, although a bit more embarrassing than doing it at home; some groups work out outside if you are too shy of being by yourself. I need to do yoga and meditate to avoid anxiety, and I feel I actually can achieve more things in the day when I practice them than if I try to save time by not practicing.

- **Be grateful**: Remember what you have achieved so far and appreciate your achievements. Writing them down can help you feel better about what you have accomplished and help you achieve more in the future.

  At times, we might think that we are not getting closer to our goals, but if we look back, we will see all the progress we have made so far. If you are just starting, then be appreciative of the opportunities ahead, known and unknown.

There are many other things to be grateful for and people to show this appreciation to who have helped you in your journey. If you cannot think of anyone, try to find people that are active and supportive, and join a community; the testing community is generally very healthy and helpful. (See *[1]* in the *Further reading* section for some suggestions on communities.)

- **Avoid comparisons**: Sometimes, we might get discouraged when we compare ourselves with other people. Try to avoid comparison. Everyone has times when they are more active or lucky, so it makes no sense to make this comparison. Everyone has different areas of focus or expertise and different styles, so you will always be able to contribute with yours.

  Remember that copying others can only bring you a certain degree of success if any. Instead, work on improving their vision and ideas, mixing different ideas together to create something new, channeling those ideas, and always giving credit to the person whom you got the idea from. It feels very good to hear that your idea inspired someone else.

Be patient and persistent during this process.

Now that we know some techniques to challenge our careers, let us discuss how to find out what to work on so that we can set our goals and activities around those projects.

## Finding hot topics

The best way for me to find inspiration about things I am doing is to keep track of new technologies and think about how to use them to improve my daily performance. You should work on what excites you and makes you want to learn more (what is *hot* for *you*) rather than something fashionable or popular. However, to understand what it is that works for you, you should first get informed.

You can try searching for trending technology topics or following newsletters or podcasts about technology and development. *[2]*

We have reviewed some *hot topics* in this book, such as *the cloud*, *AI (and other associated keywords, such as ML or deep learning)*, *data science*, and *XR (and other associated keywords, such as VR, AI, or the metaverse)*. In each related chapter, we have explored a bit more about how to learn more about the topic.

You can also find some trends related to particular technologies. For example, Cypress, **behaviour driven development** (**BDD**), and shifting left. These have had peaks of popularity in recent years.

However, you might want to specialize in a particular technology, technique, or program, whether this is related to something popular or not, which is totally fine. It may well be that you are the one who makes the thing you are working on popular because you figure out something deep, new, or useful. In the next section, we will try to identify your working style so that you can understand yourself better and focus on what works best for you.

# Finding your style – horizontal versus vertical knowledge

In *Chapter 1, Introduction – Finding Your QA Level*, we discussed the differences between horizontal and vertical scaling in testing. Your working style and knowledge can also tend to be horizontal or vertical.

For example, this book is quite horizontal. That is my style. I like knowing a little about many things but tend to get bored when I dive too deep into just one thing. While I like digging into a challenge or problem and fully understanding it and its solution, I also like variety, or, I shall say, so far, I have preferred variety with respect to picking up projects. I feel it helps my creativity because sometimes I can mix solutions that I have seen working with another technology and find new solutions that way. I prefer the overall picture so that I can think out of the box. However, this could change at any time.

Maybe you prefer to dive deep into any particular topic, say unit testing. You can become an expert on it and know everything that there is and has been in relation to the topic. That is *OK*, too; the beauty of the world is in the differences, and there is always room for everyone and every style. Just make sure you understand which one is yours.

Once you have figured out what you want to work on and you are inspired by it, the next problem is getting time to work on it.

## Finding time to work on that hot topic

Once you have something you would like to work on or a planned project, you will need to find time to work on it. In this section, we will see some of the tactics you can try to make time for your career growth project.

### *Organizing your priorities*

Let's see how to set of order of priority of your tasks and work on them effectively.

#### Planning

Planning is a good way of finding priorities. You can be as detailed as you want on a plan as long as the things you are planning for are realistic. You can add an hour a week for research and 3 hours for learning, or whichever amount works for you. Make sure you stick to the plan afterward as much as you can but do not be too hard on yourself if you can't; life is not something we can predict.

#### Learning to say no when needed

Consider whether you are the best person to take on the task, whether it is related to your goals, whether you will be able to accomplish it, and whether the company will benefit from you doing it. On the other hand, if you are new or have the time, take everything you can because it will help you learn more and grow (but always from the perspective that you will be able to accomplish it).

### Using the Pomodoro technique

A technique that works wonders for me is the **Pomodoro technique**. It consists of taking a timer to count the time you need to focus and, after that, some time for a break. For example, you can use 20 minutes of full focus and 5 minutes for a break. In the first few loops, it might be difficult to focus for 20 minutes, but after a while, it gets easier and easier, and you might want to increase it to 25. The hardest part of physically moving some big machine (for example, a heavy cart) is to get it to start to move. Mental focus works in the same way.

### Using reminders and post-it notes

Another technique is to have reminders in your calendar and as post-it notes. I like having quotes in my working environment that remind me why I do what I do.

### Identifying your distraction agents

Identify your distraction agents so that you can reduce such distractions. For example, if you are using the Pomodoro technique, you can keep your phone in a different room or inside a drawer and check it only during your breaks.

If your plan goes well, you might be able to find time where you did not have enough of it before, as you will get more efficient.

## Applying your project to your job

If it turns out that your company's needs are the same as the needs that will be sorted by working with your topic of interest, you will not likely struggle to find that time within your working hours. However, things get more complicated if your company has other priorities. You then need to figure out a way of making this topic a priority, or you need to figure out how you can reduce the load or delegate to allow you to work on it. This is, of course, taking for granted that realizing this project will help your company too.

## Automation

One way of reducing such load is to work on automation, and I don't mean just testing automation. If you figure out a way of automating as many repetitive tasks as possible, then you will get the best currency of them all, an invaluable one: **time**. You should be able to use this time to work on your topic and add the cherry on top of whatever priority you are currently working on.

By now, you should be able to know the steps to automate anything, and what sorts of things are worth automating. Try to apply this information to anything around you, as we highlighted in *Chapter 11, How to Test a Time Machine (and Other Hard-to-Test Applications)*; even if the time saved is small at first; it will accumulate and become a bigger time saving over time.

### Work on the idea in your free time

If you cannot fit this project into your job even after planning and automating, or simply, it does not fit your company's needs, you can work on this in your free time. Since this is a topic that really excites you, it will likely not be such a burden to spend some time on it after working hours. Be careful not to get burnt out, and check whether your company's policies allow you to do this kind of work in your free time. Use planning and automation for your off-work activities as well.

If you are interested in a particular technology, make sure it is the right move for your company too; otherwise, it is better to learn it in your free time.

Now that we have understood how to grow, we can start helping others grow too. Let us see how you can build the right team for the job.

# Quality growth

To create the best quality of applications, we must make sure that we have the right roles in our team. Note that I purposely did not specify the usual *right people for the job* because I feel some people might work well in different roles. Let us then review what types of roles are available in quality.

## Roles for testing

We discussed the roles for testing in the very first chapter of this book *Chapter 1, Introduction – Finding Your QA Level*. Feel free to review that chapter for more information if you do not feel comfortable with the descriptions in this chapter. In order to look for the right skills to grow within a company, it is imperative that you understand what the exact role entails. As we mentioned in *Chapter 1*, different companies define the roles differently, so job descriptions and internal descriptions are a good way of figuring out the needs for each position. If your company does not have such descriptions, you can help them create them. I have been asked many times for help defining test roles *[3]*, interviews, and even salaries. My general definitions are in *Chapter 1*, although more specifications would require understanding the exact needs of the company and the team.

Once we have defined the different roles, even if they are not called the same thing, we can look into what to do to help our team members grow.

## Helping people in your team grow

In this section, we will go into detail about helping the other team members to grow. It does not matter if you are not a manager, you can help yourself and others grow with these ideas. Different levels of experience will require different capabilities. Senior positions should be more independent and probably do not need as much help, but they still need support to allocate the required resources, networking, and feature requirements setting to achieve the planned goals.

Let us go one by one through some of the roles in the development life cycle and see how they could be assisted.

### Helping manual testers grow

Manual testers can be more efficient in their jobs if they have tools that help them achieve more in less time, such as a record-and-playback tool, for example, or automating ways they enter defects in the system.

When manual testers are in charge of the definition of the test cases, using methodologies such as BDD can help them participate early in testing discussions and guarantee there is little repetition of testing done as they will be aware of what has been automated. This technique only works when the team is involved, which means you need to spend time and resources from the other roles here, which could mean less time to work on coding. Done in any other way, it makes little sense, and it is too time consuming.

Some manual testers are interested in writing code. In those cases, you can provide courses for them to learn about it. If you are a manual tester and you want to have more knowledge in coding, you could try to get a part-time software development degree. It is not mandatory, but it would be helpful, as there are a lot of things to know to create good programs, and you can prove your passion for it. Besides, degrees provide guidelines and bases for learning more; they can help you grow into architecture since you will learn to design systems and to grow horizontally too. Again, it is not mandatory as you can get this knowledge from many places, including mentoring, reading designs and documentation, and your own experience building things and projects. Furthermore, some degrees teach things that you might never use in your working life, which might be good for some people, but others might prefer focusing on the useful parts. Whichever way you get the knowledge required to create good programs, you should definitely have passion about it, otherwise, it will just feel exhausting and frustrating to learn it. I do not recommend companies forcing manual testers learning about automation for this very same reason.

Other manual testers might be interested in growing into a management positions. The skills they might need for this will be specific to each company, but mentoring could be of help for most people considering moving into management as they will experience working with others to help them ' grow. I have found mentoring amazing, as you can reach more through your mentees than you would have done on your own. However, as of the time of this writing, I would not like to become a manager, as there are more skills needed that I do not think match my current passions. For the other skills needed other than mentoring, specifically related to business and managing resources, you could get a specific mentor that can guide you through the learnings needed and possibly some certifications that might be of help too.

If you are unclear about what to do, keep in mind that you can always try different things and that as life changes, you might also change your mind.

## Helping developers in test (SDET) grow

**Software developer engineers in test** (**SDETs**) are a rare species and dare I say it, one likely to be extinct soon too. Three Microsoft employees first coined this term in the book *How We Test Software at Microsoft [4]*, explaining it to be a developer specialized in building quality tools, in a similar way that there are front-end or back-end specialized developers. Many companies followed this approach, while others struggled with the idea, giving the position a "bad name" among developers, as sometimes they would request a good percentage of manual testing. After some time (and I presume some difficulty finding qualified developers that wanted to focus on quality), Microsoft decided to give every developer the title of *Software engineer* instead, starting the "shift-left" movement in such a way (therefore making the remaining SDETs even rarer).

SDETs exist in the dichotomy of being developers with a testing mentality. Therefore, being stuck in writing test code can be frustrating for most of them as this code is generally repetitive and not always particularly challenging. Make sure you empower them by helping them identify processes to automate and tools they could write to keep improving their programming skills, and that could benefit the rest of the team.

Allow them to work on heavy development and practice other development specialties so that they can move to that role if need be.

## Helping developers grow...with quality

Learning about quality techniques and theory can help developers grow too. As they work on different programs and they make sure they add quality, they will start writing the programs with quality in their mind. This will mean that they will slowly prevent defects while they develop the code, as they will be used to the checks that will be performed afterward and will keep those in mind at time of developing.

Some developers like building their own tools but tend to forget to do that when they are constrained to features coding. Thinking about quality and automation can help get them back on track and inspire them to create other tools that could benefit everyone in the company.

If you think a developer is losing passion in what they are doing, let them innovate with quality in mind. It is also possible that they would prefer to move to management as well.

Note that not everyone wants to transition to a new role or position, and that is totally fine. But that does not mean that you cannot still grow within your own role, constantly learning and coming up with new things. Even if you do not want to grow, I think growth is something that will happen naturally anyway, and remember you still have a lot to share with others to help them grow, even if you prefer staying in your comfort zone.

Now that we have reviewed how the different roles can grow, we will see more about transitions and career opportunities in the next section.

# Transitions and career opportunities

If you want to advance your career, besides working on projects that will help you move in that direction, you should consider preparing for interviews, marketing yourself, and working on promoting yourself to the right mediums so that you get the offers for the positions you would like. This should be done irrespective of whether you can and want to grow within your current company because it will help you understand your market value and will keep you in motion.

As mentioned in the previous section, consider having a mentor to work more specifically on the skills you will need to make the transition. Let us now see some requirements for entering the different testing related position and how you could prove your skills for it.

## *Interviewing for test-related/QA positions*

It would be impossible for me to predict what you would get asked in an interview, as they vary quite a lot from company to company. There are books written on each possible type of interview, so it would be redundant for me to go over those as well.

Generally speaking, though, manual test positions will get asked more questions about types of testing, hypothetical systems for which to create test plans, and the theory of testing.

Managerial positions will be asked about how they would deal with different people's problems, how to organize the work, and several other hypothetical cases and real scenarios that they will have experienced in their careers.

Test architects will be asked about systems and framework designs, besides other development questions. They can be assessed on the quality of the written code and design, usage of data structures and software principles, and other technologies relevant to the specific job's needs.

Developers in tests will be asked similar questions as the manual testers but add more complicated code and system design to the picture. The interview should be pretty similar to that of a developer, given I am a firm believer that testing questions should be part of all the interviews for the people involved in the development life cycle of every application, especially if your company is considering shifting left. Depending on their level of experience, they might start to get asked some of the test architect questions. Depending on the job's needs, they might get specific questions about the technologies they require for the job.

Other roles, such as solutions architect, might require specifics about dealing with customers.

If you are looking to prepare yourself for such interviews, remember that the most important part is to be honest about your CV and your skills. When the interview is related to your previous experiences, it should be easy for you to come up with examples that will showcase your abilities. Another important aspect of an interview is to be able to express yourself clearly and in enough detail. If you do not provide sufficient detail, your answer might be too vague, and your contributions hard to see. However, if you extend yourself for too long, interviews will not have enough time to cover all the interview's topic and you will fail the interview, so be specific and concise about what you want to tell them.

One exercise you might want to do is to prepare some examples of your biggest project on some cards, expressing in a summarized but clear way what was the problem that you resolved, what your exact contribution was, what the result was, and what you learned from it or how would you have done it differently. Try to use this method to explain any project or process you implemented or contributed to. The objective is to use your interview time wisely and have a mental structure that will help you get out of a *blank state*.

Besides this, if you are working towards a technical interview, then review common coding exercises to remember patterns and software principles. You should be able to resolve any of these challenges without practice, as it is meant to be part of your day-to-day job. However, I feel much more relaxed after I have practiced and proven to myself that I can do it, otherwise, I would get anxiety during the interview.

Even so, sometimes I have gotten into a *blank state* in interviews and proceeded to an absolutely terrifying *brain freeze* in which I cannot answer or perform the easiest of tasks. This can happen to anyone at any time; interviewing is a process, do not be hard on yourself if that is the case, and keep trying.

In the following section, we will see the skills that you may need if you are about to transfer roles. Check them carefully and make sure you give yourself projects and time to work on these skills so that you can be confident and relaxed if you are interviewed about them. Remember that the interview preparation should come through your work. If you pretend to know something you have not worked on or study things for an interview a month before it takes place, it might help you pass it, but it will show during your employment that you are not experienced. Therefore, do not skip the next section, and make sure you include this in your current job before searching for the change.

### Transition skills

The following table is a summary of the skills you should work on if you are planning a role transition. There is much more to each of the transitions, and it depends highly on the specifics of the company, but let me try to give you a rough idea. The vertical is the starting role, and the horizontal is the role you desire to move toward:

|  | **Manual Tester** | **Developer in Test** | **Manager** | **Test Architect** |
|---|---|---|---|---|
| Manual Tester | Nothing | Coding – specific technology and languages | People skills – organizational skills – business – test theory | Frameworks – design – developer in test skills |
| Developer in Test | Test types and theory | Nothing | People skills – organizational skills – business – test theory | Frameworks – design |

| Manager | Test types and theory | Coding – specific technology and languages | Nothing | Frameworks – design – developer in test skills |
|---|---|---|---|---|
| Test Architect | Nothing | Nothing | People skills – organization skills – business-test theory | Nothing |

We considered that the skills needed for developers and developers in test should be the same ones, especially if we are considering the *shift-left* approach. However, the type of work and working style for developers could differ from the one for developers in test. Generally, developers in test require more creativity and more responsibility in the products built, while other types of developers work on small features of a bigger product.

You might think that it should be easier for a developer in test to move to an architect position as they have been working on architecting tools from end to end; however, it really depends on the complexity of the tools you have worked on.

Besides this, if you are planning to move from SDET to SDE on a different specialization (for example, for DevOps or frontend), then consider preparing for it by working on tools that will showcase the experience in that specialization.

We have not added the hybrid new role of *test automator*, which is a manual QA tester that has some degree of knowledge about writing coding for automation, as they surely know how to grow into the role of developer in test or manual tester and they can infer the rest from the table.

If you are considering any position that we have not included here, such as security expert, system reliability engineer, or data analyst, make sure you understand what it entails. Check out different companies' role descriptions and definitions to get more information about what you would need for them. Then use the same technique as described before: find a way to include those missing skills in your current job. Your mentor or manager might help you with this (as long as the position is needed for the company). If you can't, then consider joining a course about them or working on a side project that will give you the experience (especially if you work with others that can mentor you on this in such projects).

We have mentioned the importance of mentoring a couple of times in this chapter already. In the next section, we will see this in detail.

### Mentoring plan

Having a mentor can greatly help you, as they will be able to give you specific hints to achieve your results. You can have more than one mentor and have them in the long term or short term, for general help and guidance or for a specific project.

On the other hand, mentoring also has its benefits. It is nice to be nice, but that is not the only reason you should consider mentoring others. Among the benefits of mentoring are that it may remind you of your early passion and why you love doing what you do when you work with someone starting with something. It can also help you to be more organized, as you need to be able to express what has worked out for you in a repeatable way for your mentee to understand it and use it. Mentoring helps me in these ways. You could also have more reach through your mentees than you would on your own, by influencing them or delegating tasks to them.

Most people who start with mentoring understand that it could be beneficial but struggle to put it into practice. Here are some of the things that have helped me with mentoring. Feel free to use them when you are involved in mentoring as a mentee or mentor, and consider adding them to a shared document for both people involved to contribute to it:

- **Introduction**: Even if you know the other person from before, start by explaining your experience up to now. You can have a CV walkthrough, including courses and certificates, and even share your actual CVs. That works for both parties, so it is clear what they can offer and request. Summarize this information in the session's notes.

- **Objectives**:

  - What would you like to obtain from these sessions?

  - What is your dream job? What does it look like? Be specific, including what type of company it would be (international, startup, freelance). What type of position or role is it? What would your day-to-day responsibilities be? What about your schedules, challenges, colleagues, technologies, locations, and benefits?

  - What would you like to specialize in? (Remember vertical versus horizontal.)

  - How long can you spend on growth activities outside of your work?

  - Are there any role models you would like to follow?

  - What sort of learning do you prefer (audio, video, books, tutoring, or courses)?

- **Project updates**: Specify anything that you have been working on in relation to the specific project or process you are asking for mentoring on:

  - What have we achieved so far?

  - What is to follow? (Action items.)

  - What are we struggling with?

- **Feedback on the mentoring sessions**:

  - What is going well?

  - What needs to improve?

  - Do the sessions still make sense or shall we change them somehow?

Feel free to add your own points and your specific needs.

After you have worked on yourself and created processes and projects that have helped you grow, you should work on your self-promotion. We will see some hints on this in the next section.

### Marketing yourself and self-promotion

Talking about marketing and self-promotion can be as awkward for you as it is for me, but it is nonetheless necessary. If that is the case, it could be a good opportunity to get out of your comfort zone. As much as you could be hard working, if you are unable to showcase your work and remind the key individuals in your organization of such work, you will rarely get to grow much, unfortunately.

Every organization has its own rules and promotion opportunities, but here we are not referring only to that. Even collaboration among other team members or participating in events requires a degree of showcasing your abilities.

If you keep track of your work as we recommended in the first section of this chapter (*Challenging your career*), you are halfway there. Try to share that track or keep it in a public place so that others are aware of it, can applaud it, and even find you to get more information about anything in there.

One way of sharing your achievements and helping raise awareness of other people's achievements is by having an online presence. However, make sure that you are not a slave to it; remember that it is ok to miss out. You can also join external conferences and talks, or internal events and presentations within your own company.

Sometimes it is hard for us to see the positive things that we do, as we feel that our contributions are not important. This is the case when you suffer from *impostor syndrome*, and we will review it in the next section.

## Impostor syndrome

Note that this topic deals closely with mental health and concepts such as self-esteem that might require the help of a professional therapist (which I am not). If you struggle with this, consider seeking therapy, but know that you are not alone. Keep in mind that the suggestions in this section are things that helped me but might not work for you as we are not built in the same way.

I am of the opinion that referring to this syndrome is not the best idea, as words have power, and they can make you feel like one just by saying it. However, this is an important topic, and I feel it is necessary to include it in this book. It is one of the main stoppers for your personal growth.

> Impostor syndrome
> When you do not feel good enough, or you feel different than others, or both.

Sometimes, when you feel like that, you might be right. You might not be good enough for your own standards, and you are unique in some way. Therefore, if you don't want to feel this way, make sure you have realistic standards about yourself and what you need to achieve and try not to compare yourself with others. Everyone is distinct in their own personal way and can bring something different to the table. Let us see other ways I use to deal with my impostors.

## How do I deal with my impostors?

I feel we all have multiple of these *impostors* inside of us, little voices that tell us that we are not enough in different ways. It might sound crazy, but I try to *talk* back to these voices, learn from them, and ensure that their concerns have been heard. I cannot say, therefore, that I have overcome them, but I have learned to live with them without them taking control of my life (at least, most of the time).

I remember one of my biggest fears while talking at a conference (or even writing this book) was that I was going to say something that someone else had already said. It took me a while (and some friends calling this out to me in different ways and times) to realize that even if that were the case, they would not say it in the same way. Find your voice and your style, and be kind to yourself.

When you feel that the imposter feeling is coming from outside of you, remember that most of the time this is just in your head. If the critics are still obvious, let people be wrong, there is no need to prove yourself to anybody but yourself. There will always be people who hate you and love you. Try to surround yourself with the latter, and just pick a bit of the former so that you can learn and grow from their feedback. Besides, people that underestimate you will get the biggest surprise when you finally demonstrate your skills.

Lastly, sometimes the way people act towards you just shows their own insecurities, as we are mirrors of each other. That said, be also kind to others, and use your empathy so that you don't produce impostor feelings in them.

There are times when the impostors' voices might get louder, such as when public speaking, a deadline is coming closer and you have still not completed the task, or even during a job interview. These times you might require extra effort to deal with your impostors, but every time it gets a little bit easier. Realize that nobody was born knowing everything, and failure is just a way to get closer to success, an opportunity to learn.

Be realistic with your actual qualities: if you need a pencil, a pen will not be good enough. However, the pen would be much better than the pencil under different circumstances.

Feeling you are not good enough is a great incentive to improve, so use it to your benefit.

Let us close this chapter and this book with some final thoughts.

# Closing thoughts

In this section, we will discuss some final topics that I felt were important to add to this book. All these are my personal opinions, and I have a lot of room for discussion, so let us start discussing it in a healthy way. Feel free to reach out to me online and let me know your thoughts about them.

In the next section, we will discuss ways to find goals to work on.

## Finding your why

In a book by Simon Sinek called *Find Your Why [5]*, the author starts a movement to help people find things that inspire them. I think this is also valid for the quality field.

It is important to keep your current goal in mind to focus on what is important for you, your career, and your organization. You can have more than one goal in mind. Let me give you an example.

My main goal in life is to make the world a better place, even if just a little bit. This goal is very broad, which allows me to break it down into sub-goals that are more specific and achievable. For example, writing a book that helps others could be one way to make the world a better place. There could be other ways of making the world a better place, such as donating your time, effort, or money to some organization you care for.

Let us break it down even further: the main reason for me writing this book is to help someone, even if just one person. This is a realistic and achievable goal. If my goal were to help everyone, I would likely fail and feel bad about myself for not being able to reach it.

Once you find your *why*, you should be able to come up with many ways that will work towards that why, or you will find your *what*, the project to work on that will align with such a *why*.

If this is still not enough, think about what you want to offer. Is there something, some experience, or skill you can offer to others that could help your *why*? If there is nothing, then think about the way you want to grow, the things you want to learn and achieve to reach this point where you will be able to support your *why*.

Think about what success will look like and how you will measure it when you reach that goal.

We could also discuss goals in quality and make sure your team and company are aligned with them. Why do we work on quality? Why should we care? Maybe we want to give the best experience to our users; for what purpose do we seek this? We could use the bottom-to-top approach with this. Let us be honest and reach the origin of this purpose too: is it to earn more money? Is it so we can be proud of our product instead of embarrassed by it? Is it so that our product helps our customers and does not get in their way? For their satisfaction? For our learning and expansion? To prove our value or growth? Maybe it is for several of these reasons.

Once we have a clear vision, we can break it down into smaller achievable goals as well to have a clear path for our company's quality. If this exercise is done by all the teams and roles involved in the development life cycle, then we will have an alignment about why quality is important, and that will raise awareness about it and increase the quality of our products.

Life is a process, not a destination, and as such, you will need to iterate through this several times and keep coming up with exciting stuff to work on and give to others.

One of your goals could be to work in an ethical way. Let us have some words about ethics in testing.

## Ethics in testing

Would you work for a company where you are comfortable and earn good money but that has ethical issues that you do not agree with, given that you have other options? Sometimes it is the only way to go, even if you have very few options or take a cut in your salary or benefits. However, I think it is important to be able to call out ethical reasons and be an advocate for less privileged groups.

It is not easy to change a company's or group's culture, especially when the rest of your colleagues might not seem to share your values or care to speak up. At times, it takes one person to speak up for the rest to start following too. You do not need to be extremist about it; sometimes, it is just a matter of stating a fact so that others can understand that there is an issue.

If you are a manager, make sure your employees are given power and a voice to make a change when needed. Listen to them. The world is evolving, and we should try to evolve with it.

Different fields are more prone to ethical issues. For example, in artificial intelligence, we need to make sure that our training sets are inclusive. For face or voice recognition, we should make sure we train our system so that nobody is left behind or unable to use the system. We should also make sure we are not creating an all-powerful humanity-killing machine.

Another example is that of **extended reality** (**XR**), which we have already discussed in *Chapter 10, Traveling Across Realities*: we should make sure that people feel safe in virtual environments and in their interactions with them.

We should also take care of social media and ensure we can take down a publication as quickly as possible, but also that we can re-establish it quickly if later proven to be safe.

It seems as though ethical issues are going to start gaining importance in the future of quality, especially in new fields. In the next section, we will discuss what the future of quality might look like.

## The future of quality

In this section, I will give you my feelings about the changes that the QA world will go through in the coming years based on the trends I see in the industry.

As mentioned in the last section, ethical issues will gain importance in testing and development, while for AI apps and tools, it might be irrelevant to have a separate role for testing as the training will take care of this. Besides this, tools such as GitHub's Copilot can help create testing code more easily, reducing the number of people needed to perform this task.

We can also find tools such as *Applitools* that allow for visual testing, getting smarter, and achieving more of the tasks that used to be impossible to automate before.

However, if you work as a manual QA, you should not be concerned, as there will always be room for manual testing, especially at the first stages of an application. I feel that there will either be companies or sub-teams within a company that will work on manual testing until things get automated.

I also think there will always be a need for experts in testing to make sure the best tools and techniques are being employed at the right time, especially if companies decide to shift left. This technique could be detrimental to quality if no experts advocate for and teach it, which could make companies revert the process.

However, test experts might start taking different roles other than *test expert*, such as site reliability engineers. This could potentially be an important role as we try to deploy CI/CD, and with systems built partially in the cloud, each component's quality will be guaranteed by a different team or company, as we saw in *Chapter 9*, *Having Your Head up in the Clouds*.

Users' feedback will be very important, and companies will focus on acting on such feedback as quickly as possible.

SDETs (of which there are few already) might start changing roles, and the role might fade away as we shift left. SDETs might move to more specialized roles or development.

In the next section, we will see how you can work in the future so that you keep ensuring the quality of your company, team, and work.

## The future of you

This chapter is the last one of this book, and I usually get emotional when it comes to the last chapters of books. I hope you have enjoyed this virtual trip with me so far and there is at least one thing in this book that could be of use for you, or better yet, impacted you or will make you enjoy your career more and get better at it.

Whatever step you take in your testing career, and whatever stage your application, company, or team is, utilize the skills and techniques of this book to grow to the next level. Keep growing, moving, and learning. Make sure you set high goals for yourself but that they are realistic. Break those goals down into smaller ones to make sure you will achieve them, and give them an adequate timeline.

Use the concepts in this book to experiment and learn, create, and discover something new and cool, something people would like to read and write about. I am looking forward to seeing these creations; it would be a dream come true for me if we happen to cross paths in the future and you tell me that this book helped you in any way, shape, or form.

## Summary

Funnily enough, as I am writing these lines, I am actually facing a big case of impostor's syndrome as the book is about to be done and become a reality. Writing this chapter has helped me remember why I was writing the book in the first place and inspired me to keep working hard on it. I hope this energy flows to you somehow through these pages, too, and that you enjoyed reading it as much as I enjoyed writing it.

In this chapter, we saw some tips on how to challenge your career, find topics that inspire you to get to the next level, align them to your working and learning styles so that you can build your career plan, and finally, find time to work on it. We also learned ways of helping others in your team grow, reaching even higher levels, and how to handle self-sabotage. We finalized this chapter by talking about a very important subject that must never be forgotten: your reason for doing what you do, alongside the ethics related to testing and what the future could possibly hold.

As Mark Twain said: *"The secret of getting ahead is getting started."* Go and do epic quality stuff! Best of luck, and have loads of success and fun!

## Further reading

- *[1]* Some test communities could be specific to a tool or framework, such as Selenium (`https://www.selenium.dev/support/`) or Testim (`https://www.testim.io/community/`), or could be related to a conference, such as the Test Guilds (`https://guildconferences.com/community/`), but there are also specific groups, such as the Ministry of Testing (`https://www.ministryoftesting.com/slack_invite`). All of them are great resources to network with others.

- *[2]* Examples of podcasts about technology and development, specifically related to testing: `https://testguild.com/podcasts/` (find me on episode 223: `https://testguild.com/podcast/automation/223-testing-dream-journaling-smashing-sand-castles-with-noemi-ferrera/`) and `https://podcasts.apple.com/us/podcast/ministry-of-testing/id1046372364`. Also, try searching for video feeds on different platforms about the testing world or specific tools.

- *[3]* I talk about the roles I have held here: `https://thetestlynx.wordpress.com/2017/04/27/between-two-rivers/`.

- *[4]* *How We Test Software at Microsoft* by Alan Page, Ken Johnston, and Bj Rollison.

- *[5]* *Find Your Why: A Practical Guide for Discovering Purpose for You and Your Team*. Part of: *Start with Why Series*, by Simon Sinek, David Mead, et al.

# Appendix – Self-Assessment

In this appendix, we will put together the different sections that we reviewed in *Chapter 1, Introduction – Finding Your QA Level*, so that you can get an idea of the quality level of your group or company. If there are any terms you are not sure about, or you are unfamiliar with, feel free to revisit *Chapter 1* to refresh them. Take this appendix as a guide to see whether you can incorporate testing aspects from the next level. Keep in mind that you might already be right where you need to be.

## Assessment

To measure your quality level, pick the best answer for your team and company and write down the count of each of the letters *a–d*. After that, we will do some calculations:

1. **Development process**:

   a. Waterfall

   b. Fake agile

   c. Agile

2. **Development cycle**:

   a. Test last

   b. Test at the same time

   c. Test first (TDD or maybe BDD with the test defined with behavior)

3. **Team**:

   a. QA and development teams separated

   b. Developers work closely with testing but separate CRs

   c. CRs have developers and test experts

4. **Team 2**:

   a. SDETs do most testing

   b. QAs do most testing

   c. Developers do most testing

5. **Deployment**:

   a. Require manual intervention

   b. Deployments are tested offline or at off-peak hours

   c. Fully built-in CI/CD

6. **Regression**:

   a. New features can affect old ones

   b. New features rarely affect old ones

   c. New features are handled by feature tags

7. **Type of testing**:

   a. Some testing (usually unit or integration) and the rest manual

   b. Unit tests and end-to-end or UI tests

   c. Unit tests, integration tests, and end-to-end tests

   d. As per *c*, plus security, accessibility, performance, or other types of specific tests

## Score:

Here is where we calculate the score:

Add all *a*'s and multiply them by one. Then, add all *b*'s and multiply that result by two. Add to the previous value. Then add all *c*'s together and multiply the result by five. Add this to the total.

Finally, add five points, plus 1 for each extra test type if you answered *d* in the last question.

The results of your quality level are listed here:

- If you scored less than 10, you have further to go in terms of quality. Follow the tips in *Chapter 1, Introduction – Finding Your QA Level*, and across the book to improve your system. If you have a very small user base and deal with basic information, this could be just fine.

- If you scored more than 10 but less than 35, you are doing reasonably well with your quality but could do much better. Follow the tips in the book to get to the next level. However, if you deal with critical or high-risk information or have a high number of users, you should make even more effort toward improving your quality.

- If you scored more than 35, you are above average with your quality already; we hope you find some tips in this book that you have not considered yet.

- If you scored more than 50, you probably made a mistake. Error: score overflow.

# Index

## Symbols

`Packtpub.com`

Subscribe to our online digital library for full access to over 7,000 books and videos, as well as industry leading tools to help you plan your personal development and advance your career. For more information, please visit our website.

## Why subscribe?

- Spend less time learning and more time coding with practical eBooks and Videos from over 4,000 industry professionals

- Improve your learning with Skill Plans built especially for you

- Get a free eBook or video every month

- Fully searchable for easy access to vital information

- Copy and paste, print, and bookmark content

Did you know that Packt offers eBook versions of every book published, with PDF and ePub files available? You can upgrade to the eBook version at `packtpub.com` and as a print book customer, you are entitled to a discount on the eBook copy. Get in touch with us at `customercare@packtpub.com` for more details.

At `www.packtpub.com`, you can also read a collection of free technical articles, sign up for a range of free newsletters, and receive exclusive discounts and offers on Packt books and eBooks.

# Other Books You May Enjoy

If you enjoyed this book, you may be interested in these other books by Packt:

**Test Automation Engineering Handbook**

Manikandan Sambamurthy

ISBN: 9781804615492

- Gain a solid understanding of test automation
- Understand how automation fits into a test strategy
- Explore essential design patterns for test automation
- Design and implement highly reliable automated tests
- Understand issues and pitfalls when executing test automation
- Discover the commonly used test automation tools/frameworks

**Software Test Design**

Simon Amey

ISBN: 9781804612569

- Understand how to investigate new features using exploratory testing
- Discover how to write clear, detailed feature specifi cations
- Explore systematic test techniques such as equivalence partitioning
- Understand the strengths and weaknesses of black- and white-box testing
- Recognize the importance of security, usability, and maintainability testing
- Verify application resilience by running destructive tests
- Run load and stress tests to measure system performance

## Packt is searching for authors like you

If you're interested in becoming an author for Packt, please visit `authors.packtpub.com` and apply today. We have worked with thousands of developers and tech professionals, just like you, to help them share their insight with the global tech community. You can make a general application, apply for a specific hot topic that we are recruiting an author for, or submit your own idea.

## Share Your Thoughts

Now you've finished *How to Test a Time Machine*, we'd love to hear your thoughts! Scan the QR code below to go straight to the Amazon review page for this book and share your feedback or leave a review on the site that you purchased it from.

`https://packt.link/r/1801817022`

Your review is important to us and the tech community and will help us make sure we're delivering excellent quality content.

# Download a free PDF copy of this book

Thanks for purchasing this book!

Do you like to read on the go but are unable to carry your print books everywhere?

Is your eBook purchase not compatible with the device of your choice?

Don't worry, now with every Packt book you get a DRM-free PDF version of that book at no cost.

Read anywhere, any place, on any device. Search, copy, and paste code from your favorite technical books directly into your application.

The perks don't stop there, you can get exclusive access to discounts, newsletters, and great free content in your inbox daily

Follow these simple steps to get the benefits:

1.  Scan the QR code or visit the link below

https://packt.link/free-ebook/9781801817028

2.  Submit your proof of purchase
3.  That's it! We'll send your free PDF and other benefits to your email directly

www.ingramcontent.com/pod-product-compliance
Lightning Source LLC
Chambersburg PA
CBHW062046050326
40690CB00016B/2999